Key Challenges in Geography

EUROGEO Book Series

This book series addresses relevant topics in the wide field of geography, which connects the physical, human and technological sciences to enhance teaching, research, and decision making. Geography provides answers to how aspects of these sciences are interconnected and are forming spatial patterns and processes that have impact on global, regional and local issues and thus affect present and future generations. Moreover, by dealing with places, people and cultures, Geography explores international issues ranging from physical, urban and rural environments and their evolution, to climate, pollution, development and political economy.

Key Challenges in Geography is an initiative of the European Association of Geographers (EUROGEO), an organization dealing with examining geographical issues from a European perspective, representing European Geographers working in different professional activities and at all levels of education. EUROGEO's goal and the core part of its statutory activities is to make European Geography a worldwide reference and standard. The book series serves as a platform for members of EUROGEO as well as affiliated National Geographical Associations in Europe, but is equally open to contributions from non-members.

The book series addresses topics of contemporary relevance in the wide field of geography. It has a global scope and includes contributions from a wide range of theoretical and applied geographical disciplines.

Key Challenges in Geography aims to:

- present collections of chapters on topics that reflect the significance of Geography as a discipline;
- provide disciplinary and interdisciplinary titles related to geographical, environmental, cultural, economic, political, urban and technological research with a European dimension, but not exclusive;
- deliver thought-provoking contributions related to cross-disciplinary approaches and interconnected works that explore the complex interactions among geography, technology, politics, environment and human conditions;
- publish volumes tackling urgent topics to geographers and policy makers alike;
- publish comprehensive monographs, edited volumes and textbooks refereed by European and worldwide experts specialized in the subjects and themes of the books;
- provide a forum for geographers worldwide to communicate on all aspects of research and applications of geography, with a European dimension, but not exclusive.

All books/chapters will undergo a blind review process with a minimum of two reviewers.

An author/editor questionnaire, instructions for authors and a book proposal form can be obtained by contacting the Publisher.

Jakub Szabó • Paula Puškárová
Mikuláš Černota

Governance of Social Tipping Points

Resilience of the European Union's Periphery vis-à-vis Migration, Climate Change and War

Springer

Jakub Szabó
Institute of European Studies
and International Relations
Faculty of Social and Economic Sciences
Comenius University in Bratislava
Bratislava, Slovakia

Paula Puškárová
Department of International Trade
Faculty of Commerce
University of Economics in Bratislava
Bratislava, Slovakia

Mikuláš Černota
University of Economics in Bratislava
Bratislava, Slovakia

ISSN 2522-8420 ISSN 2522-8439 (electronic)
Key Challenges in Geography
ISBN 978-3-031-47412-5 ISBN 978-3-031-47413-2 (eBook)
https://doi.org/10.1007/978-3-031-47413-2

This Springer imprint is published by the registered company Springer Nature Switzerland AG
The registered company address is: Gewerbestrasse 11, 6330 Cham, Switzerland

Paper in this product is recyclable.

Acknowledgments

The authors want to disclose following funding. This work was supported by the Slovak Research and Development Agency under the Contract no. APVV-21-0237, Ministry of Education, Science, Research and Sport of the Slovak Republic under the VEGA contract no. 1/0689/23, and European Commission's Horizon 2020 Research and Innovation Programme under Grant agreement no. 822806.

Contents

List of Figures

List of Tables

Chapter 1
Introduction

On February 28 and April 4, 2022, respectively, the Intergovernmental Panel on Climate Change (IPCC) released the second and third volumes of its Sixth Assessment Report (IPCC, 2022a, b). The panel, made up of world-renowned experts on social–ecological systems, warned against mismanaging global climatic and societal challenges, which could ultimately lead to irreversible consequences and potential tipping points with cascading effects beyond system boundaries. The adverse threats of climate change plaguing the global community are almost too numerous to be counted, ranging from melting ice and permafrost to wildfires, groundwater salinity, floods, and sea level rises. Not only are these challenges dangerous in and of themselves, but they can also affect entire ecosystems, species, and key infrastructures of cities and human settlements and translate into the degradation of economic activities and livelihoods, health risks and diseases, food, and water security, not to omit the gradual dismantling of political and legal systems, cultural values, and trust in international relations. According to the IPCC (2022a: 52), climate hazards (e.g., droughts, tropical storms and hurricanes, heavy rains, and floods) associated with extreme events have both direct (e.g., shelter destruction) and indirect (e.g., rural income losses) causes of involuntary migration and displacement. Climate change is expected to initially displace large numbers of people from the countryside to the cities, and when life in the urban centers becomes unsustainable, many will continue migrating elsewhere (Rigaud et al., 2018). According to Balsari et al. (2020), it is now recognized that climate change is a threat multiplier that will accelerate the decision to migrate and will disproportionately affect already vulnerable communities.

Forecasts for worst-case scenarios ("business-as-usual" approach) show that almost a third of the world's population will eventually live in uninhabitable climate zones. Combined with context-specific social, political, geopolitical, and economic drivers, environmental degradation is displacing vulnerable populations. Migration and conflicts related to climate change and resulting agricultural problems are increasingly present, for example, in the semi-arid Sahel, which stretches across

J. Szabó et al., *Governance of Social Tipping Points*, Key Challenges in Geography,
https://doi.org/10.1007/978-3-031-47413-2_1

Africa from Ethiopia to Senegal. In the Sahel, agriculture and livestock farming are becoming increasingly difficult for smallholders due to the lack of water in the Sahara, contributing to conflicts in Nigeria (Akinyemi & Olaniyan, 2017), Uganda (Branch, 2018), Sudan (Mazo, 2009), and Kenya (Parenti, 2011). Many people, faced with an unbearable climate and food insecurity leading to violent conflicts, have left their homes and migrated to urban centers, neighboring African countries, and some even across the Sahara and the Mediterranean into Europe. In addition, in the Middle East and North Africa (MENA) region, climate change along with water scarcity and droughts in the East and Northeast, mismanagement of watersheds (unmodernized irrigation systems), and consequent impacts on regional hydrology had a profound impact on Syria's civil war that eventually led to the displacement of large populations from their ancestral lands to urban centers, including the seat of the Assad government in Damascus, further contributing to food insecurity and sky-rocketing unemployment (Gleick, 2014). The underlying political grievances in these urban areas, already hamstrung by 1.5 million Iraqi refugees from previous years, were additionally aggravated by this disenfranchised population, turning the spring 2011 street protests into an all-out civil war that killed around 500,000 people, displaced another 10–11 million people, and ultimately made sure 5.5 million more decided to seek refugee (Balsari et al., 2020). From today's perspective, we can see how climatic and socioeconomic stressors triggered a turning point and fueled a massive exodus of fleeing refugees and migrants from the Sahel and MENA region to Europe with over 1 million migrants and 1.2 million asylum applications in 2015 alone (Dagi, 2017).

The summer 2021 heatwave, when Europe was hit hard by a series of extreme climate events – devastating floods in Germany and Belgium, or wildfires in Turkey, Greece, and France – was a stark reminder that human impact on the Earth's climate has not only indirect consequences in triggering climate-related wars in distant lands and subsequent waves of migrants and refugee seekers to continental Europe but also how such changes affect the everyday lives of average Europeans. One aspect of the European environment that is particularly sensitive to climate change is the forests, mainly due to the long life spans of trees not allowing for a quick adaptation to structural changes (Lindner et al., 2010). It is well known that changes in climate affect abiotic disturbances, such as the frequency and intensity of fires and storms, as well as biotic disturbances, such as outbreaks of pests and diseases. Several authors have argued that the occurrence of exothermic organisms (e.g., herbivorous insects) and fungal diseases is strongly modified by climate conditions (e.g., Desprez-Loustau et al., 2007). This, in turn, may affect the ability of forests to provide economic, social, and other ecological services, particularly in the areas of timber production, non-timber forest products (e.g., berries and mushrooms), carbon sequestration, biodiversity, recreation, and conservation (Lindner et al., 2010). What a climate change–induced epidemic of exothermic organisms can do might be exemplified by bark beetle outbreaks, such as those occurring in eastern European countries, for instance, in Tatra Mountain in the Slovak Republic. It has long been emphasized that eastern European countries are not adequately prepared for intense bark beetle outbreaks in the wake of climate change, which may lead to significant social and political consequences (Hlásny et al., 2021). Furthermore, it has also been pointed out that once a tipping point is crossed and epidemics occur, bark beetle outbreaks are associated with spillover disturbances to other ecosystem areas such as water extraction and soil stability, soil nutrients, carbon extraction and organic carbon stocks, and eventually even air quality degradation (e.g., Strzyżowski et al., 2018). Thus, it is important to emphasize that beyond "traditional" channels through which climate change can affect the social–ecological systems, forestry-related disasters and the ensuing pest epidemics can severely affect a wide range of human activities dependent on forests, such as timber production and tourism.

Just 4 days before the release of the second volume of the IPCC's Sixth Assessment Report, geopolitical tensions on the European Union's (EU) eastern border have crossed a tipping point and entered into a new systemic state. Ongoing geopolitical tensions between Russia and Ukraine, which escalated with the annexation of Crimea in 2014 and the conflict with Kremlin-backed separatist militias in eastern Ukraine, have escalated into a full-blown war with the invasion of Ukraine by Russian troops on February 24, 2022. Similar to the climate-induced displacement of entire populations in the Sahel and MENA regions, Ukrainian refugee flows did not begin with Russian aggression but can be seen as a continuation of a lack of political, economic, and human security in Ukraine (Lloyd & Sirkeci, 2022). Over the past three decades following the Ukrainian independence in 1991, the Ukrainian population has harbored persistent migration aspirations driven by economic deprivation, widespread corruption, and state capture by the powerful group of economic elites (Mol et al., 2017). Even before 2014, it was estimated that 6 million Ukrainians were working and living abroad (Vollmer, 2016). The fall of the regime of pro-Russian President Yanukovych in 2013, the emergence of separatist movements in eastern Ukraine, and the annexation of Crimea in 2014 naturally changed the composition of migration drivers significantly, from a country suffering from economic insecurity to a one ravaged by war. In the ensuing conflict, 1.7 million Ukrainians were internally displaced, and Ukraine became the third largest asylum seeker in the EU. Prior to the 2022 invasion, it was relatively difficult to predict the extent of the invasion and subsequent migration burdens on EU countries; albeit, it was reported even before February that Ukrainian billionaires were fleeing the country in anticipation of the war (Lloyd & Sirkeci, 2022). As soon as the first military boot crossed the Russian–Ukrainian border, the system reached its tipping point with disastrous consequences. One of the immediate consequences of this conflict was the mass exodus of Ukrainians seeking temporary shelter in neighboring EU countries. According to the United Nations High Commissioner for Refugees (UNHCR), over 7.8 million refugees from Ukraine were registered throughout the whole of Europe in the aftermath of the invasion, and 6.2 million Ukrainians remain internally displaced due to the war (UNHCR Data Portal). Of course, migration has not been the only consequence of this geopolitical turning point. Another consequence is the EU's continued dependence on Russian fossil fuels and the weaponization of Russian gas supplies to Europe in the context of Russia's military expansion in Ukraine. Although this interdependence was largely seen as a positive example of EU–Russia cooperation in the early 2000s, its negative aspects emerged after the gas transit crisis in Ukraine in 2006 and 2009, not to mention the outbreak of the Ukraine crisis in 2014 and the subsequent deterioration in EU–Russia relations (Siddi, 2022: 238). Despite Russian-backed separatist tendencies in eastern Ukraine and the annexation of Crimea, energy trade between the EU and Russia has indeed experienced "anni mirabiles" in the post-2014 period (Henderson & Sharples, 2018). The invasion of Ukraine by Russian forces in 2022 highlighted the fragility of the EU's dependence on Russian energy sources. Between February 2021 and 2022, the price of natural gas quadrupled from €20 to €80/MWh, driving up electricity prices while Gazprom cut off gas supplies to Poland, Bulgaria, and Finland. Needless to say, as energy prices rise, European industry is threatened with economic collapse, a large number of households are driven into poverty, and the embargo on imports of Russian oil further divides the EU, turning European public opinion against the measures taken against the war aggressor (Osička & Černoch, 2022).

Whether it is massive refugee and migration flows into Europe, a pest epidemic in eastern European forests, or geopolitical tensions on the EU's eastern border and the accompanying refugee and energy crisis, all three cases have two major things in common. First, they all seem to represent tipping points that once get crossed, become irreversible, spread, and contaminate other areas of social life, Second, they all seem to affect the EU's eastern and southern periphery countries the most (see also Alessi et al., 2020). As a result, two clear puzzles arise from these three cases in our opinion. First, it is not clear how the (geographical) periphery of the EU manages to cope with tipping points having significant societal impacts and how resilient the EU periphery is against nonlinear and accelerating social–ecological changes resulting from migration, climate change, and war. So far, to the best of our knowledge, no study has complexly and comparatively tried to assess pivotal determinants distinguishing a successful governance of such social–ecological disturbances from a failed one within the EU peripheral context. Second, it also remains to be seen how the governance of abrupt nonlinear social–ecological changes, or rather resilience against them, is spatially and geographically conditioned given the core-periphery EU dichotomy and how the resilience of the peripheral region influences the resilience of the EU as a whole. Peripheral regions are generally considered the weakest links in the overall resilience of any community; however, they also tend to be the most overlooked given their inaccessible location. There are already some indications that peripheral EU regions, especially those with external borders, tend to be less resilient than regions with no national border or where these borders are internal to the EU (e.g., Healy & Bristow, 2019); however, robust multiple case study research regarding the governance of social tipping points was not conducted as far as we know. Hence, this monograph aims to examine the issue of tipping point governance and resilience to abrupt, nonlinear social–ecological change in the peripheral context of the EU. This becomes increasingly important as the concept of resilience now seems to be high on the EU's list of priorities, becoming a "compass" for EU policies (EC, 2020), justified not only by the impact of climate change but also by the emergence of artificial intelligence and security threats (e.g., EUGS, 2016). To tackle the two aforementioned puzzles, this monograph adopts the multilevel stakeholders' agency approach to examine how the various state and non-state actors (transnational, private, and local) have managed to navigate social tipping points with different triggers (migration, climate change, and geopolitics) in the eastern and southern periphery of the EU, in countries such as Greece, Slovakia, and Poland. We are particularly interested in assessing the role of the state in managing social tipping points and how state resilience is related to the resilience of non-state actors within the multilevel governance framework that is constrained by peripheral geographical position. This will allow us to identify pivotal determinants of resilience against abrupt changes in the EU's periphery (Ingalls & Mansfield, 2017).

This book thus proceeds as follows: Before addressing the specific aspects of social tipping points on the eastern and southern borders of the EU, we first need to provide a general theoretical overview of what tipping points actually represent and whether there are differences in the perception of tipping points across different academic fields, namely in environmental and social sciences. In doing so, the first section focuses primarily on the literature review of tipping points in the environmental (earth and ecological) and social sciences with possible overlaps and intersections between these two fields. It is argued that despite the sometimes-antagonistic presentation of tipping points in the environmental and social sciences, the most appropriate strategy for our use is to adopt an interdisciplinary approach to tipping points and position them within the social–ecological system. Thereafter, we operate with tipping points as the concept of thresholds at which an abrupt change triggers a nonlinear transformation in the social component of the social–ecological system; the transformation is driven by a self-reinforcing feedback mechanism and ultimately leads to a qualitatively different state of the social system. Furthermore, it also tends to be of limited reversibility (hysteresis). Then, it can be argued the social component within social–ecological systems that changes the most is the migration and displacement of people affected by a tipping point, although this is definitely not the only social component that changes as we also have to acknowledge the importance of (local) economy and politics.

The subsequent chapter addresses the conceptual framework of state governance, state capacity, and resilience in the face of social tipping points and their mutual intersections. After adopting a state-centric view of governance and resilience, we argue that it is the state that remains at the forefront of addressing the impacts and adverse consequences of social tipping points. In our framework, governance represents a dynamic process of the general exercise of state authority when confronted with certain material conditions, such as a social tipping point. It consists of multiple steps, such as goal selection, goal alignment and coordination, goal implementation, feedback, and accountability. It is argued that the most important aspect that determines the overall success of the state in addressing social tipping points is the state's ability to impose its will, i.e., state capacity, which consists of five critical subcomponents – administrative, legal, infrastructural, fiscal, and military capacity. Resilience, the central concept in this book, represents the flip side of the state's capacity: the state's ability to withstand disruptions due to social tipping points and its capability to reorganize or adapt to them. Although the state and its resilience represent the first line of defense when a social tipping point occurs, some of its effects can penetrate through the umbrella of state resilience. In this case, non-state actors (transnational, private, and local) enter the scene to complement the state in combating the effects of the social tipping point. It is assumed that non-state actors can ultimately increase not only the overall societal capacity and resilience but also the capacity and resilience of the state itself, and, therefore, it is in the interest of the state to allocate resources to enhance non-state capacity and resilience as well.

However, the stakeholder's agency as a prerequisite for social tipping points governance alone is not sufficient. Given the scope of the monograph primarily deals with the EU's eastern and southern periphery, Sect. 4 also takes into account spatial and geographical determinants of tipping points governance, capacity, and resilience. It is demonstrated that there is a discrepancy between how the core of the EU on one hand, namely the western and northern EU countries, and the periphery of the EU on the other, especially the eastern and southern periphery, varies in their capacity and resilience against social tipping points. It is also shown that there is a clear association between the capacity to be resilient and the level of economic development across the EU member states to the detriment of its periphery. Furthermore, building on the literature on regional economic resilience, major determinants of geographically conditioned resilience are presented. In particular, compositional, collective, and contextual factors are identified. Compositional factors are closely related to structural composition, collective factors refer to networks and dependency beyond the stakeholders, whereas contextual factors put an emphasis on the importance of institutions. A combination of all these factors contributes to the relatively low levels of resilience against social tipping points in the EU's eastern and southern periphery.

Once we deconstruct the theoretical aspects of tipping points, introduce public policy concepts with a focus on governance, capacity, and resilience in the context of the multilevel governance system, and uncover the spatial and geographical determinants of the distribution of capacity and resilience across the EU member states, Sect. 5 summarizes the theoretical concepts and provides a conceptual framework of the governance of social tipping points in the EU periphery. Based on the deductive-nomological conceptual framework, an empirical strategy and methodology are devised. We then formulate the research question and research goals and state the hypotheses. To answer them, we employ a multi-case research design with three cases of potential social tipping points in the EU periphery, each with a different trigger: migration, climate change, and geopolitics.

In the ensuing section, we present the case study reports with the aim of identifying whether a social tipping point occurred in that case, what was the state capacity and resilience against that potential tipping point, and what was the respective capacity and resilience of the non-state actors. First, for migration-induced tipping points, we focus on the burning of the Moria camp on the Greek island of Lesbos, which can be indirectly attributed to the civil wars in Syria and sub-Saharan Africa and the subsequent migration flows to continental Europe. Second, we examine the 2004 windstorm in the Slovak Tatra Mountain as an example of a climate-induced social tipping point that eventually led to an outbreak of bark beetles due to the lack of state resilience. Third, we turn to the case of geopolitical tensions on the eastern periphery of the EU that led to the invasion of Ukraine by the Russian army. Against the backdrop of this geopolitical turning point, we examine how different actors contributed to the management of the refugee crisis on the Polish–Ukrainian border and what interactions existed between the state and non-state actors' resilience. Based on the cross-case conclusions stemming from our case study reports, we reflect our theory (conceptual framework) and develop policy recommendations in the last section. We conclude that while state capacity and resilience are critical in addressing social tipping points, the collaboration of non-state actors (transnational, private, and local) with their respective capacities and resilience can mitigate some of the consequences resulting from social tipping points and help the state to prevent yet another tipping point from occurring. The coalition of state and non-state actors is crucial in this regard, as demonstrated by the example of Poland's management of the refugee crisis on its border with Ukraine. In this context, it is crucial that the state strengthens not only its capacity and resilience but also the capacity and resilience of its non-state actors.

The final section contains several policy recommendations for dealing with social tipping points from the perspective of the EU periphery. First, we recommend not focusing on governing social tipping points per se, but rather on governing their side effects. Second, we recommend strengthening state capacity in all its aspects – administration, law, infrastructure, taxation, and military – as these were identified as key drivers of state resilience. Third, as we demonstrate the importance of the resilience of non-state actors, we recommend that the state and society invest more resources in strengthening the resilience of transnational, private, and local non-state actors as it also enhances their respective capacities and resilience. Fourth, it is clear the EU periphery is highly dependent on the EU, and this dependence appears to be mutual. To make the EU periphery more resilient, we suggest pooling more resources into building a more resilient EU. It is argued that the stronger the EU, the better it is for all of its member states including the eastern and southern periphery in terms of resilience building. Fifth, it is suggested that not only is the periphery dependent on the well-functioning EU, but the EU's resilience also requires a fortified periphery given the deep interconnectedness between the geographical core and periphery. Lastly, we conclude by pointing out the benefits of social–ecological inter-dynamics for studying social tipping points and recommend that policymakers and key stakeholders change their methodological approach from purely social to social–ecological systems.

References

Akinyemi, E. T., & Olaniyan, A. (2017). Nigeria: Climate war. Migratory adaptation and farmer-herder conflicts. *Conflict Studies Quarterly, 21*, 3–21.

Alessi, L., et al. (2020). The resilience of EU member states to the financial and economic crisis. *Social Indicators Research, 148*, 569–598.

Balsari, S., Dresser, C., & Leaning, J. (2020). Climate change, migration, and civil strife. *Current Environmental Health Reports, 7*(4), 404–414.
Branch, A. (2018). From disaster to devastation: Drought as war in northern Uganda. *Disasters, 42*(S2), 306–327.
Dagi, D. (2017). Refugee crisis in Europe (2015-2016): The clash of intergovernmental and supranational perspectives. *International Journal of Social Sciences, 6*(1), 1–8.
Desprez-Loustau, M., et al. (2007). Simulating the effects of a climate-change scenario on the geographical range and activity of forest-pathogenic fungi. *Canadian Journal of Plant Pathology, 29*, 101–120.
EC. (2020). *Strategic foresight report*. European Commission. https://commission.europa.eu/strategy-and-policy/strategic-planning/strategic-foresight/2020-strategic-foresight-report_en
EUGS. (2016). *Shared vision, common action: A stronger Europe*. A global strategy for the European Union's Foreign and Security Policy. European Union Global Strategy. https://www.eeas.europa.eu/sites/default/files/eugs_review_web_0.pdf
Gleick, P. H. (2014). Water, drought, climate change, and conflict in Syria. *Water, Climate, and Society, 6*(3), 331–340.
Healy, A., & Bristow, G. (2019). Borderlines: Economic resilience on the European Union' eastern periphery. In G. Rouet & G. C. Pascariu (Eds.), *Resilience and the EU's eastern neighborhood countries: From theoretical concepts to a normative agency*. Palgrave Macmillan.
Henderson, J., & Sharples, J. (2018). Gazprom in Europe – Two "Anni Mirabiles" but can it continue? *Oxford Energy Insight, 2008*(29).
Hlásny, T., et al. (2021). Bark beetle outbreaks in Europe: State of knowledge and ways forward for management. *Current Forestry Reports, 7*, 138–168.
Ingalls, M. L., & Mansfield, D. (2017). Resilience at the periphery: Insurgence, agency and social-ecological change under armed conflict. *Geoforum, 84*, 126–127.
IPCC. (2022a). *Climate Change 2022: Impacts, adaptation and vulnerability*. Working Group II contribution to the IPCC Sixth Assessment report of the intergovernmental panel on climate change. https://www.ipcc.ch/report/ar6/wg2/downloads/report/IPCC_AR6_WGII_FullReport.pdf
IPCC. (2022b). *Climate Change 2022: Mitigation of climate change*. Working Group III contribution to the IPCC Sixth Assessment report of the intergovernmental panel on climate change. https://www.ipcc.ch/report/ar6/wg3/downloads/report/IPCC_AR6_WGIII_Full_Report.pdf
Lindner, M., et al. (2010). Climate change impacts, adaptive capacity, and vulnerability of European forest ecosystems. *Forest Ecology and Management, 259*, 698–709.
Lloyd, A. T., & Sirkeci, I. (2022). A long-term view of refugee flows from Ukraine: War insecurities, and migration. *Migration Letters, 19*(4), 523–535.
Mazo, J. (2009). Darfur: The first modern climate-change conflict. *The Adelphi Papers, 49*(409), 73–86.
Mol, C., et al. (2017). Migration aspirations and migration cultures: A case study of Ukrainian migration towards the European Union. *Population, Space and Place, 24*(5), 2131.
Osička, J., & Černoch, F. (2022). European energy politics after Ukraine: The road ahead. *Energy & Social Science, 91*, 102757.
Parenti, C. (2011). *Tropic of chaos. Climate change and the new geography of violence*. Nation Books.
Rigaud, K., et al. (2018). *Groundswell: Preparing for internal climate migration*. World Bank.
Siddi, M. (2022). EU-Russia energy relations. In M. Knodt & J. Kemmerzell (Eds.), *Handbook of energy governance in Europe* (pp. 237–262). Springer.
Strzyżowski, D., Fidelus-Orzechowska, J., & Żelazny, M. (2018). Sediment transport by uprooting in the forested part of the Tatra Mountains, southern Poland. *Catena, 160*, 329–338.
Vollmer, B. (2016). *Ukrainian migration and the European Union – Dynamics, subjectivity, and politics*. Palgrave Macmillan.

Chapter 2
Tipping Points: A Survey of the Literature

The use of the tipping point framework has been on the rise, especially since the early 2000s (Nes et al., 2016; Milkoreit et al., 2018). However, despite its widespread use, the concept is still relatively vague, especially within the social sciences. As a methodological framework, tipping points originated in the natural sciences, particularly chemistry (Hoadley, 1884) and mathematics (Poincaré, 1885). Poincaré, for instance, mathematically described a point in the function of a system at which a small change in a parameter can cause a sudden change in the behavior of the dynamical system, leading to the appearance of a new solution with different properties. Poincaré referred to this as bifurcation theory (Poincaré, 1885), and it remains widely used in mathematics, physics, and the science of complex systems.

Subsequently, the concept was rediscovered in the environmental and social sciences. Grodzins (1957) was one of the first to introduce the concept into social sciences, while Schelling famously formalized and popularized it (Schelling, 1971, 1972, 1978). Following Schelling's seminal work, various terms describing abrupt changes in qualitative states of social and environmental systems emerged in the late 1990s, preceding the spread of the term tipping point in the social science literature by nearly a decade (Milkoreit et al., 2018). Prior to that, various synonyms were used, such as threshold, regime change, and critical transition. Even Gladwell (2000: 17), author of the magnum opus on tipping points, defines the tipping point as "the moment of critical mass, the threshold, the boiling point." Similar to today's conceptualization of tipping points, there was no consensus on what these concepts represent. For instance, while the economic sociologist Granovetter (1978) understood the concept of threshold as a vague change in people's behavior, institutionalists such as Gould and Eldredge (1993) viewed threshold as a precisely delineated boundary of an equilibrium. Although all of the aforementioned works were undeniably influential in the social sciences, it was not until climate change science rediscovered the tipping point framework in the analysis of complex systems that its academic use skyrocketed (Rusill & Nyssa, 2009; Kopp et al., 2016; van der Hel et al., 2018).

Before we get into the nuances of the tipping points in different academic disciplines, let us briefly summarize the underlying logic behind the concept of tipping points. Tipping points depict a situation, where once a certain threshold is passed, the dynamics of the change accelerate and cause a paradigmatic shift from one state to a completely new, qualitatively different state (Scheffer et al., 2009). The tipping

J. Szabó et al., *Governance of Social Tipping Points*, Key Challenges in Geography,
https://doi.org/10.1007/978-3-031-47413-2_2

points, thereby, represent unanticipated nonlinear changes in the observed systems and can be used in two distinct meanings, portraying either a regime shift and a qualitative change in the structure and functioning of the social system (Folke et al., 2004; Scheffer, 2009) or a regime transformation, involving a nonlinear systemic change (Gunderson & Holling, 2002; Olsson et al., 2014).

Naturally, there are different understandings and nuances to this complex concept, especially across different fields. Since the early 2010s, we have registered a growing divergence between the usage of the tipping point concept in ecological systems (Lade, 2013; Broderstad & Eythórsson, 2014; Serrao-Neumann, 2016), on the one hand, and social tipping points (STPs), framework predominant in social sciences, on the other (Grimm & Schneider, 2011; Sims et al., 2016). As Milkoreit et al. (2018) rightly point out, there are noticeable differences between ecological and social systems; therefore, their conceptualizations of the term will be slightly different. On the whole, ecological and social systems are not the same ontological entities, even though they complement each other in certain aspects. In order to be able to capture all the nuances of the tipping point framework holistically, we review tipping point literature in both fields, environmental and earth systems as well as social sciences, compare them, and explore their mutual interconnectedness. Ultimately, we argue that the best way to study the governance of tipping points and potential resilience against them we shall consider the social and ecological systems as complementary and we, therefore, focus on the social tipping points within the social–ecological systems (SESs). This way, we will be able to conceptualize resilience against tipping points arising from the ecological sphere but having social side effects and vice versa.

2.1 Tipping Points in Environmental Sciences

Contemporary public discourse frequently discusses topics of the fragility of the environment, ranging from the Amazon rainforest to Arctic permafrost, pointing to possibly irreversible human-made damage to the environment. As a result of this, the earth system sciences and ecology literature are relatively ample in dealing with multi-stability systems and non-convexities, as a proxy for the tipping points (Dasgupta & Mäler, 2003; Groffman et al., 2006; Mumby et al., 2007; Lenton et al., 2008; Biggs et al., 2009). Scholars studying social–ecological systems adopted the concept of tipping points in different variances, ranging from social–ecological regime shifts (Lade, 2013) and tipping points in social–ecological systems (Broderstad & Eythórsson, 2014) to social–ecological tipping points (Serrao-Neumann, 2016). The majority of these papers deal with environmental sustainability and strategies to avoid fatal ecological disasters.

A significant part of the literature in natural sciences also employs the tipping point framework to determine the threshold of carbon dioxide concentration in the atmosphere, which cannot be crossed without irreversible damage. Lemoine and Traeger (2014), for instance, managed to identify the existence of a precise threshold of carbon dioxide concentration that would lead to triggering the tipping point. Chavas et al. (2016) constructed a threshold quantile autoregressive model to characterize anomalies in atmospheric carbon dioxide concentrations. The authors examine the paleoclimate data going back 400,000 years, using the Vostok ice data series, looking for evidence of irreversibility and evidence of tipping points. Their results indicate local instability in carbon dioxide, the ipso facto existence of tipping points in the carbon dioxide concentration dynamic.

Environmental scholars very often also deal with the tipping points in the context of the ice sheet dynamics in the Arctic. Lindsay and Zhang (2005), observing the satellite pictures of summer Arctic Sea ice, noticed record lows for ice thickness in 2002–2005. They argue that during the late 1980s and early 1990s, mankind could have passed the tipping point during which the ice-ocean system began to enter a new era of thinning ice and increasing summer open water. Holland et al. (2006) employed seven projections from the Community Climate System Model and found that drastic reductions in ice extent trends will lead to near-ice-free conditions by 2040. However, they argue that observations are still too imprecise to assess whether the tipping point has already been reached. Winton (2006) claims that when the polar temperature rises above the −5 °C threshold, his models indicate an abrupt elimination of Arctic ice.

Numerous authors include the human factor in the analysis concerning the ecological tipping points, subtextually linking environmental and social dynamic systems. Lenton et al. (2008: 1786), who define the tipping point as the "critical point at which the future state of the system is qualitatively altered," include non-climatic variables into their model for explaining and predicting tipping points, such as human activity causing large-scale changes and public awareness of such changes. In their 2016 paper, Lemoine and Traeger take a look at greenhouse gas emissions from the policymakers' perspective. The authors model rational agents deciding on greenhouse gas emissions against Knightian uncertainty, processing the unexperienced chance of irreversibly tipping the planet into a less favorable climate system (Lemoine & Traeger, 2016). Also, Heutel et al. (2016) examine optimal climate policy from the perspective of policymakers, in the presence of climate tipping points defined as irreversible disturbances of the carbon climate system.

Other papers in this field assess various mitigation or adaptation strategies for the crossing of the environmental tipping point threshold. Tsur and Zemel (2016) examine the trade-offs between mitigation efforts, lowering the chance of a loss, and adaption actions, lowering the realized loss, in climate policies reacting to the crossing of the environmental threshold. So far, it has been well established that risks to biodiversity are endogenous (Shogren, 2000), and that humankind can invest in either mitigation to lower the likelihood of passing the tipping point (Horan et al., 2011), or in adaptation, once the tipping point is crossed (Knowler & Barbier, 2005; Fenichel & Horan, 2016). In certain models, both self-protection and self-insurance (mitigation and adoption) can be achieved simultaneously (Leung et al., 2002).

In terms of deterring the probability of crossing the tipping point, Baggio and Fackler (2016) investigate the possibility of imposing restrictions on the control set for ecological management of the renewable resource, a fishery, under regime-shifting dynamics. However, their research focuses on the impact of reversible regime switching on optimal harvesting policy, not on the irreversible qualitative shift between the two systems. Nævdal (2016), on the contrary, assesses how the marginal value of a resource after a catastrophe, once the tipping point is reached, affects optimal management before the catastrophe. Nævdal analyzes the effect of ex-post expectations on the current decision-making. Nkuiya and Costello (2016) examine how the possibility of an environmental regime shift affects the optimal emission policy, while Finnoff et al. (2016), very similarly, investigate the dynamics between the ex-post outcomes and the ex-ante invasive species risk management problem. The authors conclude that "an ex-ante system that is convex and uniquely stable without invasion risk may become non-convex and multi-stable in the presence of endogenous invasion risks and ex post-multi-stability (Finnoff et al. 2016: 114)."

Taking a look at the usage of tipping points in environmental studies and ecology, we come to the conclusion that in the majority of cases, the tipping point framework is referred to as a nonlinear irreversible change in system dynamics. This conclusion can be supported, for instance, by the German Advisory Council on Global Change, which perceives the tipping point as an irreversible system-changing process occurring once a critical threshold is crossed, in which case the changes are almost impossible to bring under control afterward (Schubert et al., 2008). Examples of such events can be the Arctic ice sheet, which could start to melt away uncontrollably if global warming exceeds a critical threshold, or the Amazon rainforest, which could be irretrievably eradicated if a certain number of trees are cut down. Furthermore, in the case of environmental and earth system science, we observe that the environmental tipping point thresholds are usually precisely quantifiable, with environmental scholars providing relatively exact estimates based on sophisticated models (see, e.g., Winton, 2006).

2.2 Tipping Points in Social Sciences

In the previous part, we demonstrated a relatively frequent usage of the tipping point framework in environmental and earth system literature. Even though the tipping point framework remains widely used in scholarly works in environmental studies (Milkoreit et al., 2018), more and more social scientists have started to incorporate this approach in the study of nonlinear social changes as well in the last couple of years. The usage of the tipping points in social sciences ranges from political science (Grodzins, 1957; Schelling, 1971) and sociology (Wolf, 1963) to economics (Tirole, 1993; Dasgupta & Ray, 1986; Dasgupta, 1997; Azariadis & Drazen, 1990; Diamond, 1982). As we have already mentioned in the introduction, social scientists very often adopt the term "social tipping points", to set their approach apart from environmentalists and climatologists (Skrimshire, 2008).

Following Grimm and Schneider (2011: 3), STP can be defined as "constellations where the social fabric of a country breaks apart and where a country embarks on a course of dramatic, but not always violent change." It is a point at which "the system shifts abruptly from one state to another" (Scheffer et al., 2009: 53). In social sciences, social shifts that are investigated by political scientists, sociologists, or economists deal with the transition of a social system (political system, society, or economy) from state *a* to state *b*, whereas the two are qualitatively different. Before we dive into the tipping points from the point of view of social sciences, it is important to emphasize that different social systems function on different levels. These systemic shifts occur usually on the societal level in sociology (*integrated - > segregated society*), on the state level in political sciences (*democracy - > dictatorship*), or on the economy level in economics (*conjuncture - > recession*).

Grodzins (1957) and Schelling (1971) were among the first to pioneer the tipping point approach in social sciences, investigating metropolitan segregation and concentration of the Blacks in the United States. Social researchers keep relying on Shelling's (1971, 1972, 1978) framework of tipping points in society even today. In his model, Schelling (1971) notices that it takes only a marginal increase in the number of residents with different ethnic backgrounds to incentivize the flight of other ethnic groups from the neighborhood. A couple of years later, Granovetter (1978) investigated group motives to take on collective action. He argues that the key concept is that of "threshold," described as the number or proportion of others who must make one decision before a given actor does so. Also in sociology, Huckfeldt (1980) examines the social classes in urban politics and the tipping points of the in-group population density. His findings indicate that in some cases when a tipping point is present, socially overwhelmed individuals react even in a conflicted manner to a group other than their own. The doyen of the social tipping point framework, Gladwell (2000), focuses on so-called social epidemics, and unexpected inflation of various social trends, where the tipping point represents a shift between irrelevance and spread of the social epidemics. Brock (2006) investigates the tipping points within peer-group pressure on policymakers.

In the field of political science, Skrimshire (2008) analyzes tipping points in the context of political movements in the United Kingdom, while Nathan (2013) assesses the regime stability of the Chinese communist rule using the tipping point framework. In the case of public and health policies, Sims et al. (2016) look at disease spread tipping points for evaluating the timing of public health interventions in response to epidemics, while Berry and Finnoff (2016) examine investment decisions concerning the pandemic risk to avoid disease spread tipping point. Quite important in this regard are interdisciplinary studies inquiring about so-called fragile statehood, combining political science and international relations approach. Fragile states, or failing states, lack Weberian state capacity and three basic attributes: effectiveness, authority, and legitimacy. A fragile state is not capable of sustaining the monopoly of the use of force and providing functioning social order and basic public goods (Mata & Ziaja, 2009). Yet, it is quite difficult to determine the tipping point when a state becomes failing or failed. Some political scientists developed a list of features to determine the fragile statehood status, the tipping point between two qualitatively different states, but it remains a relatively vague approximation (Bates, 2008; Goldstone et al., 2010).

When it comes to the tipping points and social sciences, they are used mostly in the field of economics (e.g., Schelling, 1978). It has been increasingly observed in economics that incremental changes, under specific circumstances, can trigger abrupt systemic shifts, at both the micro and macro levels. At the micro level, Sims et al. (2016) use the example of the price crossing a certain threshold triggering the adoption of a new technology or entry (exit) in (from) a market. At the macro level, the tipping points usually arise when different behaviors start reinforcing once a threshold is exceeded. The "domino effect" metaphor might be suitable in the aggregate level context. As an example, we can use the bankruptcy of Lehman Brothers on September 15, 2008, bringing about the Global Financial Crisis (Ivashina & Scharfstein, 2010).

To mention a few noteworthy economics articles employing the tipping points, Dasgupta and Mäler (2003) and Deissenberg et al. (2005) analyze threshold behavior, the existence of non-convexities, in dynamic economic systems with multiple equilibria. Sims et al. (2016) contend that tipping points in these dynamic economic systems arise when variables get through a certain threshold and this crossing triggers a shift from one outcome to another. Similar threshold behaviors have also been examined in modern growth models (e.g., Skiba, 1978; Jones, 1998), models of industrial organization (Tirole, 1993), models of poverty traps (Dasgupta & Ray, 1986; Dasgupta, 1997), economic development (Azariadis & Drazen, 1990), labor market (Diamond, 1982), models of externalities (Starrett, 1972), and in trade and other models of spatial economics (Krugman, 1991; Fugita et al., 1999). Wood et al. (2016) utilize an evolutionary game theory model featuring heterogeneous populations of agents and natural resource-specific variables with the computational power of agent-based modeling to examine the tipping point on the oil market from the era of Seven Sisters to OPEC (Organization of the Petroleum Exporting Countries). Similar agent-based models are used in identifying tipping points in macroeconomic models quite often (e.g., Gualdi et al., 2015; Lengnick, 2013).

Relatively precise determination of the tipping point threshold can be found in macroeconomics, fiscal policy to be more precise, focusing on finding a threshold in public debt sustainability. Reinhart and Rogoff's (2010) paper, drawing on evidence from 44 developed and developing economies, finds out that a threshold of 90% central government debt to GDP is a threshold of sustainability for the real economy. Even though the authors did not use the term STP themselves, Pozen (2010) labels Reinhart and Rogoff's threshold as a tipping point. Caner et al. (2010) also analyze thresholds in long-term average public debt to GDP ratio and its potential impact on long-term GDP growth. Contrary to other social science approaches, the tipping point, in this case, is relatively precisely quantifiable.

To summarize our findings from the tipping points in social sciences, we observe that in contrast to social–ecological tipping points (e.g., Lade, 2013; Broderstad & Eythórsson, 2014; Serrao-Neumann, 2016), social scientists do not assume that STPs need to be irreversible (Scheffer et al., 2009). On the contrary, democracies can "easily" shift back from the nondemocratic state, and the economy almost always bounces back from the recession. Furthermore, an STP does not necessarily have to lead to a negative outcome. For instance, people in post-communist countries seem to be glad, for the most part, to shift from a nondemocratic communist regime to a capitalist democracy. Some scholars even argue that STP can represent a potential solution to certain malign phenomena (Mathias et al., 2020; Milkoreit, 2023). Another important difference is pointed out by Grimm and Schneider (2011), who argue that it is by definition more difficult to define the time dimension of a sudden social shift, in comparison to delimiting the threshold for the carbon dioxide concentration, for instance. They mention the cases of the collapse of the Communist Czechoslovak regime in November 1989 or the genocide in Rwanda in April 1994, where no one was able to foresee these sudden systemic changes.

2.3 Tipping Points in Social–Ecological Systems

Reviewing the literature on the tipping points in environmental and social sciences, we managed to identify three structural differences between the two – threshold accuracy, (ir)reversibility, and outcomes. In the case of environmental sciences, models of tipping points are constructed relatively rigidly, with precisely quantified thresholds (Winton, 2006; Nævdal, 2016). Once the threshold is crossed in environmental systems, the change spins out of control leading to an irreversible process. For instance, if the Arctic ice melts down and the Amazon forests get deforested, there seems to be no coming back from this qualitative system change, at least in the short to medium run. Furthermore, the environmental tipping points tend to always lead to a negative outcome. STPs, on the contrary, are more loosely defined, so the tipping point thresholds between two qualitatively different systemic states are imprecise and often just qualitatively described. There are few exceptions though, even in social sciences, most notably in economics and finance (Reinhart & Rogoff 2010). Most importantly, STPs are reversible for the most part and do not necessarily lead to a worse outcome. For instance, the tipping point between non-democracy and democracy is relatively loosely defined, can be easily reversed, and represents a more desirable outcome.

In recent years, research on social–ecological systems (SESs), the intersection between social and natural systems, started to emerge. According to Milkoreit et al., "the SES represents an inherently linked system, with multiple feedbacks and interdependencies between the ecological and the social system components" (Milkoreit et al., 2018). We also agree with the authors that the separation of the ecological and social systems would be utterly counterproductive, and although such a mixture tends to involve the crossing of some ontological boundaries, the SESs are governed by very similar laws or principles. A similar line of argumentation was also adopted by Cote and Nightingale (2012) who contend that social and ecological system dynamics are essentially similar and criticize that some aspects of the ecology systems grown in isolation from critical social science literature. The SES as an intersection between the social and ecological systems was elaborated on by numerous authors. Lade (2013), for instance, examine the behavior of the people who interact with the ecological systems and conclude that social–ecological systems can display regime shifts that are absent from the ecological systems that are studied in isolation. Reyers et al.'s (2018: 279) research argues that SES research helps to understand how failures from conventional approaches and values emerge from ignoring or simplifying complex and dynamic social–ecological relationships. Mathias et al. (2020) analyze transition pathways and the potential outcomes that the social and ecological tipping points may lead to.

Instead of focusing solely on the social systems, we will take into consideration the SES as a whole or a social system that is in some meaningful way linked to an ecological system change (that does not exclude social determinants or causes of the change). This way, the STP "framework provides an important theoretical opening to explore the conditions for both speedy and deep, systemic changes. As such, the concept is a useful expansion of the social science toolbox" (Milkoreit, 2023), mainly with regard to the resilience in the European Union (EU)'s periphery. This, however, has affected our conceptualization of STP. We already explained that the literature on the tipping points in social and climate change sciences differs. Drawing from Milkoreit et al.'s (2018) frequency analysis across multiple disciplines, we might be able to detect a minimum set of necessary conditions to identify a tipping point within the SES: (1) *multiple stable states* that imply a certain magnitude of change and a structural reconfiguration of the system, (2) *abruptness* (or nonlinearity or disproportionality between cause and effect), (3) *feedbacks* as system-internal drivers of change the two system states as well as state stabilizers, and finally (4) *irreversibility* or more likely *hysteresis* (limited irreversibility). Subsequently, the tipping points with regard to the SES can be defined "as the point or threshold at which small quantitative changes in the system trigger a nonlinear change process that is driven by system-internal feedback mechanisms and inevitably leads to a qualitatively different state of the system, which is often irreversible. This new state can be distinguished from the original by its fundamentally altered (positive and negative) state-stabilizing feedbacks" (Milkoreit et al., 2018: 9). *Social tipping point* for the SES can be alternatively defined as the point or threshold at which "a small quantitative change inevitably triggers a nonlinear change in the social component of the SES, driven by self-reinforcing positive feedback mechanisms, that inevitably and often irreversibly lead to a qualitatively different state of the social system. Due to the interconnectedness between social and ecological system components, crossing a social (or ecological) tipping point leads to a qualitatively different SES, which is characterized by a different set of stabilizing positive and negative feedbacks" (Milkoreit et al., 2018: 10). Meeting all the four criteria (multiple states, abruptness, feedbacks, and limited irreversibility), we will adopt this definition of STP (Fig. 2.1).

One of the examples of the surrounding effects and potential feedbacks of STPs is migration. Unfortunately, the literature regarding STPs and migration is almost nonexistent. Usually, migration appears in the literature just as a side effect of global

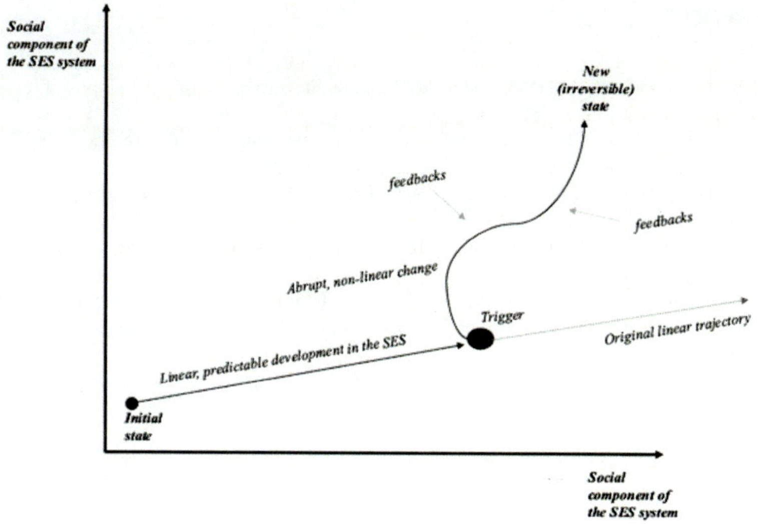

Source: Own elaboration.

Fig. 2.1 Social tipping points in social–ecological systems. (*Source:* Own elaboration)

warming, environmental disruption, or other socio-political upheavals in the articles dealing with the tipping points. Migration is, however, often the major link between the environmental and social tipping points. Missirian and Schlenker (2017), for instance, investigate how the impact of weather variations in 103 countries could translate into asylum applications to the EU. The paper concludes that temperature deviating from the moderate optimum around 20 °C increases asylum seekers' application in a nonlinear fashion, with both hotter and colder temperatures increasing migration flows. Extrapolating these trends into the future, asylum applications are expected to increase by the end of the century, on average, by 28% (98,000 asylum applications per year) under scenario 4.5 and by 188% (660,000 additional applications per year) under 8.5 scenario in climate models of the NASA NEX-GDDP. Schubert et al. (2008) presuppose that the collapse of the Amazon rainforest could fundamentally alter agricultural production in Latin America, incurring incalculable economic costs and triggering large-scale migration. Similar overlaps from climate change to global warming-induced migration were presented by Henry et al. (2004), Sinsha and Cropper (2013), the whole edition of Piguet and Laczko (2014), Bohra-Mishra et al. (2014), Rai et al. (2016), Gray and Wise (2016), and Cattaneo and Peri (2016). In spite of its relatively limited reach within STP and SES literature, migration can turn out to be a useful proxy to study EU's resilience and governance of STP since during the most abrupt changes involving qualitative regime changes, it was precisely the migration that affected the societies the most. As a result, two of our case studies deal directly with migration.

References

Azariadis, C., & Drazen, A. (1990). Threshold externalities in economic development. *Quarterly Journal of Economics, 105*(2), 501–526.

Baggio, M., & Fackler, P. L. (2016). Optimal management with reversible regime shifts. *Journal of Economic Behaviour & Organization, 132*(B), 124–136.

Bates, R. H. (2008). *When things fall apart: State failure in late-century Africa*. Cambridge University Press.

Berry, K., & Finnoff, D. (2016). Choosing between adaptation and prevention with an increasing probability of a pandemic. *Journal of Economic Behavior & Organization, 132*(B), 177–192.

Biggs, R., Carpenter, S. R., & Brock, W. A. (2009). Turning back from the bring: Detecting an impending regime shift in time to avert it. *Proceedings from the US National Academy of Sciences, 106*(3), 826–831.

Bohra-Mishra, P., Oppenheimer, M., & Hsiang, S. M. (2014). Nonlinear permanent migration response to climatic variations but minimal response to disasters. *Proceedings from the US Academy of Sciences, 111*, 9780–9785.

Brock, W. A. (2006). Tipping points, abrupt opinion changes, and punctuated policy change. In R. Repetto (Ed.), *Punctuated equilibrium and the dynamics of U.S. Environmental Policy*. Yale University Press.

Broderstad, E. G., & Eythórsson, E. (2014). Resilient communities? Collapse and recovery of a social-ecological system in Arctic Norway. *Ecology and Society, 19*(3), 1–10.

Caner, M., Grennes, T., & Koehler-Geib, F. (2010). *Finding the tipping point: When Sovereign debt turns bad* (WB policy research working paper 5391). https://documents1.worldbank.org/curated/en/509771468337915456/pdf/WPS5391.pdf

Cattaneo, C., & Peri, G. (2016). The migration response to increasing temperatures. *Journal of Development Economics, 122*(c), 127–146.

Chavas, J. P., Grainger, C., & Hudson, N. (2016). How should economists model climate? Tipping points and nonlinear dynamics of carbon dioxide concentrations. *Journal of Economic Behaviour & Organization, 132*(B), 19–45.

Cote, M., & Nightingale, A. J. (2012). Resilience thinking meets social theory? Situating social change in social-ecological systems. *Progress in Human Geography, 36*(4), 475–489.

Dasgupta, P. (1997). Nutritional status, the capacity for work and poverty traps. *Journal of Econometrics, 77*(1), 5–38.

Dasgupta, P., & Mäler, K. G. (2003). The economics of non-convex ecosystems: Introduction. *Environmental Resources Economics, 26*, 499–525.

Dasgupta, P., & Ray, D. (1986). Inequality as a determinant of malnutrition and unemployment, theory. *Economic Journal, 96*(4), 1011–1034.

Deissenberg, C., Dawid, H., & Sevcik, P. (2005). Cheap talk, gullibility, and welfare in an environmental taxation game. In A. Haurie & G. Zaccour (Eds.), *Dynamic games: Theory and applications* (pp. 175–192). Springer.

Diamond, P. A. (1982). Aggregate demand management in search equilibrium. *Journal of Political Economy, 90*(5), 881–894.

Fenichel, E. P., & Horan, R. D. (2016). Tinberg and tipping points: Could some thresholds be policy-induced? *Journal of Economic Behaviour & Organization, 132*(B), 137–152.

Finnoff, D., et al. (2016). Natural vs anthropogenic risk reduction: Facing invasion risks involving multi-stable outcomes. *Journal of Economic Behaviour & Organization, 132*(B), 113–123.

Folke, C., et al. (2004). Regime shifts, resilience, and biodiversity in ecosystem management. *Annual Review of Ecology Evolution and Systematics, 35*(1), 557–581.

Fugita, M., Krugman, P., & Venables, A. (1999). *The spatial economy: Cities, regions, and international trade*. MIT Press.

Gladwell, M. (2000). *The tipping point*. Little, Brown.

Goldstone, J. A., et al. (2010). A global model for forecasting political instability. *American Journal of Political Science, 54*(1), 190–208.

Gould, S. J., & Eldredge, N. (1993). Punctuated equilibrium comes of age. *Nature, 366*, 223–227.

Granovetter, M. (1978). Threshold models of collective behavior. *American Journal of Sociology, 83*, 1420–1443.

Gray, C., & Wise, E. (2016). Country-specific effects of climate variability on human migration. *Climate Change, 135*, 555–568.

Grimm, S., & Schneider, G. (2011). *Predicting social tipping points: Current research and the way forward.* German Development Institute, Discussion paper 8/2011.

Grodzins, M. (1957). Metropolitan segregation. *Scientific America, 197*(4), 33–41.

Groffman, P. M., et al. (2006). Ecological thresholds: The key to successful environmental management of an important concept with no practical application? *Ecosystems, 9*, 1–13.

Gualdi, S., et al. (2015). Tipping points in macroeconomic agent-based models. *Journal of Economics Dynamics & Control, 50*, 29–61.

Gunderson, L. H., & Holling, C. S. (2002). *Panarchy: Understanding transformations in human and natural systems*. Island Press.

Henry, S., Schoumaker, B., & Beauchemin, C. (2004). The impact of rainfall on the first out-migration: A multi-level event-history analysis in Burkina Faso. *Population and Environment, 25*, 423–460.

Heutel, G., Moreno-Cruz, J., & Shaegh, S. (2016). Climate tipping points and solar geoengineering. *Journal of Economic Behaviour & Organization, 132*(B), 19–45.

Hoadley, J. C. (1884). A tilting water meter for purposes of experiment. *Journal of the Franklin Institute, 117*(4), 273–278.

Holland, M. M., Bitz, C. M., & Tremblay, B. (2006). Future abrupt reductions in the summer Artic Sea ice. *Geophysical Research Letters, 33*(23), 1–5.

Horan, D. R., et al. (2011). Managing ecological thresholds in coupled environmental-human systems. *Proceedings of the National Academy of Sciences, 108*(18), 7333–7338. https://doi.org/10.1073/pnas.1005431108

Huckfeldt, R. R. (1980). Variable responses to neighborhood social contexts: Assimilation, conflict, and tipping points. *Political Behaviour, 2*(3), 231–257.

Ivashina, V., & Scharfstein, D. (2010). Bank lending during the financial crisis of 2008. *Journal of Financial Economics, 97*, 319–338.

Jones, C. I. (1998). *Introduction to economic growth*. W.W. Norton.

Knowler, D., & Barbier, E. (2005). Importing exotic plants and the risk of invasion: Are market-based instruments adequate? *Ecological Economics, 52*, 341–354.

Kopp, E. R., et al. (2016). Tipping elements and climate-economic shocks: Pathways toward integrated assessment. *Earth's Future, 4*(8), 346–372.

Krugman, P. (1991). History versus expectations. *Quarterly Journal of Economics, 106*(2), 651–667.

Lade, et al. (2013). Regime shifts in a social-ecological system. *Theoretical Ecology, 6*, 359–372.

Lemoine, D., & Traeger, C. P. (2014). Watch you step: Optimal policy in a tipping climate. *American Journal of Economic Policy, 6*(1), 137–166.

Lemoine, D., & Traeger, C. P. (2016). Ambiguous tipping points. *Journal of Economic Behaviour & Organization, 132*(B), 5–18.

Lengnick, M. (2013). Agent-based macroeconomics: A baseline model. *Journal of Economic Behaviour & Organisation, 86*, 102–120.

Lenton, T. M., et al. (2008). Tipping elements in the earth's climate system. *Proceedings of the US National Academy of Sciences, 105*(6), 1786–1793.

Leung, B., et al. (2002). An ounce of prevention or a pound of cure: Bioeconomic risk analysis of invasive species. *Proceedings of the Royal Society, Biology Science, 269*, 2407–2413.

Lindsay, R. W., & Zhang, J. (2005). The thinning of Arctic Sea ice, 1988–2003: Have we passed a tipping point? *Journal of Climate, 18*(22), 4879–4894.

Mata, J., & Ziaja, S. (2009). *User's guide on measuring fragility*. German Development Institute/ UN Development Program.

Mathias, J. D., et al. (2020). Exploring non-linear transition pathways in social-ecological systems. *Scientific Reports, 10*(4136).
Milkoreit, M. (2023). Social tipping points everywhere? Patterns and risks of overuse. *WIREs Climate Change, 14*(2).
Milkoreit, M., et al. (2018). Defining tipping points for social-ecological systems scholarship – An interdisciplinary literature review. *Environmental Research Letters, 13*, 1–13.
Missirian, A., & Schlenker, W. (2017). Asylum applications respond to temperature fluctuations. *Science, 358*(6370), 1610–1614.
Mumby, P. J., Hastings, A., & Edwards, H. J. (2007). Thresholds and the resilience of Caribbean coral reefs. *Nature, 450*(1), 98–101.
Nævdal, E. (2016). Catastrophes and ex post shadow prices – How the value of the last fish in a lake is infinity and why we should not care (much). *Journal of Economic Behaviour & Organization, 132*(B), 153–160.
Nathan, J. A. (2013). China at the tipping point? Foreseeing the unforeseeable. *Journal of Democracy, 24*(1), 20–25.
Nes, E. H., et al. (2016). What do you mean, tipping point? *Trends in Ecology & Evolution, 31*(12), 902–904.
Nkuiya, B., & Costello, C. (2016). Pollution control under a possible future shift in environmental preferences. *Journal of Economic Behaviour & Organization, 132*(B), 193–205.
Olsson, P., Galaz, V., & Boonstra, W. J. (2014). Sustainability transformations: A resilience perspective. *Ecology and Society, 19*(4).
Piguet, E., & Laczko, F. (2014). *People on the move in a changing climate. The regional impact of environmental change on migration*. Springer.
Poincaré, H. (1885). Sur l'équilibre d'une masse fluide animée d'un mouvement de rotation. *Acta Mathematica, 7*, 259–380.
Pozen, R. (2010). The US Public debt hits its tipping point. *The Boston Globe*. http://archive.boston.com/bostonglobe/editorial_opinion/oped/articles/2010/02/23/the_us_public_debt_hits_its_tipping_point/
Rai, R., et al. (2016). Climate variability and international migration: The importance of the agriculture linkage. *Journal of Environmental Economics and Management, 79*, 135–151.
Reinhart, C., & Rogoff, K. (2010). Growth in a time of debt. *American Economic Review, Papers & Proceedings, 100*(May 2010), 573–578.
Reyers, B., et al. (2018). Social-ecological systems insights for navigating the dynamics of anthropocene. *Annual Review of Environment and Resources, 43*, 267–289.
Rusill, C., & Nyssa, Z. (2009). The tipping point trend in climate change communication. *Global Environmental Change, 19*(3), 336–344.
Scheffer, M. (2009). *Critical transitions in nature and society*. Princeton University Press.
Scheffer, M., et al. (2009). Early-warning signals for critical transitions. *Nature, 461*(3), 53–59.
Schelling, T. C. (1971). Dynamic models of segregation. *Journal of Mathematical Sociology, 1*(2), 143–186.
Schelling, T. C. (1972). The process of residential segregation: Neighborhood tipping. In A. Pascal (Ed.), *Racial discrimination in economic life*. Lexington Books.
Schelling, T. C. (1978). *Micromotives and Macrobehaviour*. W. W. Norton.
Schubert, R., et al. (2008). *Climate change as a security risk*. Earthscan.
Serrao-Neumann. (2016). Marine governance to avoid tipping points: Can we adapt the adaptability envelope? *Marine Policy, 65*, 56–67.
Shogren, J. (2000). Risk reduction strategies against the explosive invader. In C. Perrings, M. Williamson, & S. Dalmazzone (Eds.), *The economics of biological invasions* (pp. 56–69). Edward Elgar.
Sims, C., Finnoff, D., & O'Regan, S. M. (2016). Public control of rational and unpredictable epidemics. *Journal of Economic Behaviour & Organization, 132*(B), 161–176.
Sinsha, P., & Cropper, M. L. (2013). *The value of climate amenities: Evidence from SU migration decision* (NBER working paper series, no. 18756).

Skiba, A. (1978). Optimal growth with a convex-concave production function. *Econometrica, 46*(3), 527–539.

Skrimshire, S. (2008). Approaching the tipping point climate risks, faith and political action. *European Journal of Science and Technology, 4*(2), 9–22.

Starrett, D. A. (1972). Fundamental noncovexities in the theory of externalities. *Journal of Economic Theory, 4*, 180–199.

Tirole, J. (1993). *The theory of industrial organization*. MIT Press.

Tsur, Y., & Zemel, A. (2016). Policy tradeoffs under risk of abrupt climate change. *Journal of Economic Behaviour & Organization, 132*(B), 46–55.

Van der Hel, S., Hellsten, I., & Steen, G. (2018). Tipping points and climate change: Metaphor between science and the media. *Environmental Communication, 12*(5), 605–620.

Winton, M. (2006). Does the Arctic Sea ice have a tipping point? *Geophysical Research Letters, 33*(23), 1–5.

Wolf, E. P. (1963). The tipping-point in radically changing neighbourhoods. *Journal of American Institute of Planners, 29*(3), 217–222.

Wood, D. A., Mason, C. F., & Finnoff, D. (2016). OPEC, the seven sisters, and oil market dominance: An evolutionary game theory and agent-based modelling approach. *Journal of Economic Behaviour & Organization, 132*(B), 66–78.

Chapter 3
Governance of Social Tipping Points and Resilience: A Stakeholder Agency Perspective

One does not have to be a social or environmental scientist to realize that contemporary Europe faces multiple challenges, be it migration, geopolitics, or climate change, and these challenges might, in turn, trigger irreversible systemic changes with subsequent spillover effects. In this context, it is of utmost importance to comprehend the concept of governance determining the resilience of states, local communities, as well as transnational complexes, especially within the context of the European Union (EU) periphery. Nowadays, governance tends to be an all-encompassing umbrella concept for a wide variety of social phenomena (Rhodes, 1997), and this is the reason why we aim to bring some clarification to this concept with regard to resilience vis-à-vis social tipping points. At the broadest level, governance is closely connected to the concept of state capacity, state strength, and governing capacity. This chapter, therefore, attempts to make a theoretical link between the concept of governance, state capacity, and resilience, all within the context of social tipping points, that will allow us to study resilience from the multilevel governance perspective. This chapter is structured as follows: First, we lay down the theoretical aspects of governance and its different components and aspects. Second, as governance remains a state-centric concept in our opinion, we ought to invest some time in linking governance with the concept of state capacity, state strength, and how these concepts relate to the concept of resilience. Third, we cannot omit that there are also other actors involved when it comes to governance, and we, therefore, ramify the concept of governance from solely state level to local, private, and transnational actors as well, arguing that the local and transnational stakeholders are often crucial in determining the overall success of resilience when the tipping points emerge. Stakeholders' agency perspective focusing on the dichotomy between the state and non-state actors with their respective capacities and resilience provides a suitable starting point upon which the conceptual framework of governance of social tipping points within the geographical constraints imposed upon the EU's periphery can be built.

J. Szabó et al., *Governance of Social Tipping Points*, Key Challenges in Geography,
https://doi.org/10.1007/978-3-031-47413-2_3

3.1 Governance and the State

By focusing on the concept of governance, on how the public sector, in conjunction with other stakeholders, such as private sector actors, transnational actors, or local communities and nongovernmental organizations (NGOs), cooperatively conduct collective actions, we might be capable of estimating, at least partially, direction and control over the society, the economy, and possibly even the resilience against unanticipated tipping points. Governance, a word closely related to the word government, can be, according to Peters (2012), likened to the process of steering a boat. As modern societies become too complex, requiring collective choices about a range of issues that cannot be addressed adequately by individual action, governance represents the captain of the boat, steering the direction of the society facing dilemmas. The concept of government in the sense of "piloting," "steering," or "directing" the society has at least four meanings in the literature: a structure, a process, a mechanism, and a strategy (e.g., Héritier & Rhodes, 2011; Pierre & Peters, 2020). As a structure, governance represents the architecture of formal and informal institutions; as a process, governance represents the dynamics and steering function involved in decision-making; as a mechanism, governance signifies institutional procedures of decision-making, compliance, and control; and as a strategy, governance stands for the actors' efforts to govern and their preferences regarding the design of institutions (Levi-Faur, 2012). Although most governance literature focuses on governance as structure, as a system of rules of governing (Enderlein et al., 2010), we believe that conceptualization of governance in this sense does not capture the potential of the governance in dealing with the systemic changes resulting from social tipping points and subsequent resilience against them. In our opinion, governance should be understood more dynamically, as a process of an ongoing process of steering or enhancing the institutional capacity to steer and coordinate (Pierre & Peters, 2020). In compliance with Peters (2012), we can identify four activities for successful governance as a process (Fig. 3.1). First, governance as a process requires some information regarding the goal toward which the society is being steered, be it performed by the state actors alone or in collaboration with other non-state actors. The *goal selection* is, therefore, crucial across all levels of the systems. Second, as the various actors within the state all have their own goals and intentions, effective governance assumes sorting priorities and coordinating the actions taken based on those priorities, which requires *goal reconciliation and coordination*. Third, once the goal is settled and various priorities within governance are sorted, reconciliated, and coordinated, governance subsequently requires *implementation*, which is, arguably, the most pivotal aspect of the whole governance process. Lastly, all the actors involved in the process need to learn from their mistakes and accomplishments, both for improving the quality of decisions and for improving democratic accountability (Peters & Pierre, 2006). Therefore, some well-developed methods of *feedback and accountability* must be part of the governance process, especially within democratic societies. These functions are fundamental to the process of governance and can be elaborated further by also taking into

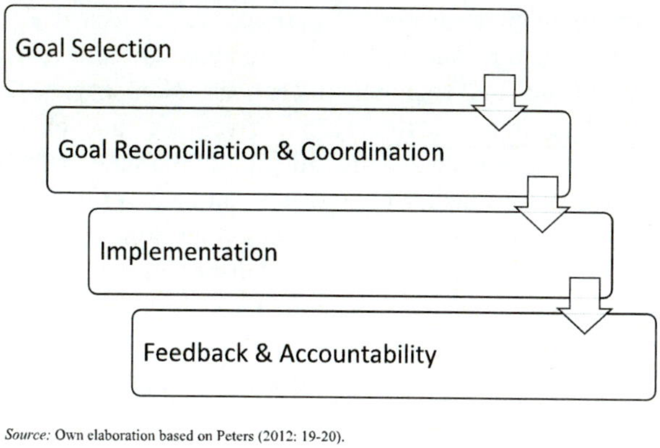

Fig. 3.1 Governance as a functionalist argument. (*Source*: Own elaboration based on Peters (2012: 19–20))

consideration other functions, such as decision-making, resource mobilization, implementation, or adjudication (Peters, 2012).

As long as it was established that governance represents a process, rather than an institutional structure, and it requires accomplishments of certain functions to be coherent, we can adopt the definition of governance as "the general exercise of authority" over the society to tackle its obstacles (Michalski et al., 2001: 9), in a Weberian sense, where authority represents institutions, both state and non-state. These institutions subsequently select goals and reconcile and coordinate those goals with subsequent implementation, whereas the implementation shall always be followed with feedbacks and accountability (Lynn, 2012). Furthermore, it is important to realize that the agenda of order, efficiency, and legitimacy within the society should always be viewed "within the state, by the state, without the state, and beyond the state" (Levi-Faur, 2012).

There are various concepts theorizing the relationship between the state and governance, ranging from the "hollowing out of the state" due to the gradual shift from government to governance (Rhodes, 1994), shift from "big government to big governance" due to the expansion of the regulatory activities of the state (Levi-Faur, 2012) to the "state-centered" governance (Pierre & Peters, 2020). In spite of the partial shift and transformation of the organization of the state in recent years, the erosion of its capacities and the onset of non-state actors in the governance process, we still agree with Peters (2012) that effective governance, at least in contemporary societies, remains a state-centric concept, which is based on the primary involvement of state actors. Pierre and Peters (2020) argue that although governance is accompanied by evolving relationships between the state and society and currently relies on less coercive policy instruments, the state remains the center of political power, making priorities, defining policy goals, and ultimately implementing them. Furthermore, most emerging forms of non-state governance depart from a model of democratic government and within the liberal-democratic theory lack accountability. Offe (2009) also argues that while non-state actors are being recruited for the fulfillment of certain public tasks through appropriate means and according to their specific competencies and resources, the state remains the sole subject responsible for regulatory oversight and economic incentives. Non-state actors are merely *licensed* to privately exercise previously exclusive state functions. This tendency represents more of a "state-organized unburdening of the state," a transfer of certain competencies of the state, whereas the transfer can be reversed by the decision of the state at any time. The state, therefore, legally speaking, is still endowed with *Kompetenz-Kompetenz* (e.g., Kelemen, 2016). Furthermore, the state-centered governance approach denotes a relatively high autonomy of the state, while the state is independent of other actors.

Hence, in our conceptualization, the state sector remains the principal source of governance mainly due to four reasons. First, due to the position of the state as the primary source of law, where constitutions represent the tip of the legal system, distributing the powers among various state institutions and organs (Kelsen, 2008). More importantly, procedures of administrative law within and among state institutions enable the state to make decisions even when they face conflicts, which is generally not the case in other mechanisms involving non-state actors (Vrabko et al., 2019). Second, as a holder of the monopoly of the legitimate use of force within a given territory, only the states are capable of imposing collective action over the whole society. Additionally, Weber also emphasizes the administrative staff – the bureaucracy – oriented to the enforcement and realization of this order (Weber, 1968). Third, the state represents an administrative and legal order claiming to bide authority not only over the state members and its citizens but over all actions that take place within its area of jurisdiction. Lastly, the state is the only subject regulating the competition for political offices and selection of the bearers of rulership according to established rules, securing, in turn, feedback and accountability as one of the fundamental features of governance (Dusza, 1989).

3.2 From Governance to State Capacity

Since we established that governance remains a state-centric concept, with the majority of the main features of governance having their roots within the state, we must not omit the concept of *state capacity*, determining the probability of the attainment of stated objects and ultimately the magnitude of its resilience. Consider the World Bank's definition of governance as "the manner in which power is exercised in the management of a country's economic and social resources for development" (WB, 1992). Such a definition of governance implicitly includes the power of the state and the quality of its institutions and policies. Referring back to the conceptualization of governance in Fig. 3.1, where governance is understood as a process by which a problem that emerged in a polity is identified and targeted, basically a set of institutions by which authority is exercised (Kaufmann et al., 1999), then governance without a proper capacity of an actor executing it becomes an empty shell. Besides goal formulation, a subsequent important aspect of governance is to sustain coordination and coherence among a wide variety of actors with differing goals, such as political actors and institutions, corporate interests, and civil society or transnational organizations (Peters, 2012). However, neither the goal selections nor the goal coordination would be sufficient without the *capacity* of the state to implement sound policies. In this regard, we must keep in mind that the concept of state capacity is closely related to that of governance; it is its element.

At the broadest level, the state capacity entails the ability of the state to create and, more importantly, to maintain order over a delineated territory, allowing the state to enact measures to protect its sovereignty such as administrating legal justice, raising taxes, and conducting independent foreign policy (Matthews, 2012). The state capacity in its narrower form represents a process of attaining and executing stated objects of the state (Zafarullah & Rahman, 2008). According to Wang (2003), state capacity stems from the state's power to monopolize the legitimate use of force along with the potential capacity to extract necessary resources from its society, mobilize consent as well as to regulate society and the economy. In a nutshell, the state "should have the capacity to formulate coherent, plausible and broadly responsible policy guidelines to attain its goals. It needs to exhibit political will in the enactment of rules and their proper implementation" (Zafarullah & Rahman, 2008: 741). With regard to our state-centric and procedural conceptualization of governance, we will not make a mistake if we assume a widely shared Skocpole's notion of state capacity as "the ability of the state to implement official goals, especially over the actual or potential opposition of powerful social groups" (Skocpole, 1985: 5). State's capacity is thus inherently tied to the state's autonomy in policy implementation. Even based on Skocpole's conceptualization, identifying the extent of the state capacity is not that straightforward, since it is not obvious how many functions the state should perform. The range of opinions may swing from one extreme to the other (e.g., Stigler & Samuelson, 1968; Stiglitz et al., 1989). Thus, in the subsequent section, selected state competencies are laid down, which represent the core of the state capacity in our opinion.

3.2.1 State Capacity Components

Although far from an exhaustive account, in line with Savoia and Sen (2012), we can map state capacities according to the function the state should perform: administrative, legal, infrastructural, fiscal, and military (Table 3.1). First, bureaucratic and administrative capacity represents a fraction of the state capacity that is responsible for keeping the state running and functioning in its day-to-day activities. It is ideally portrayed as a competent bureaucratic apparatus able to design and implement policies. Policies, as envisaged by the state, cannot be fulfilled without an obedient flock of bureaucrats, and employees of the state prepared to perform all the tasks entrusted to them. The bureaucrats conduct a wide variety of functions, ranging from issuing construction permits to running population records to healthcare surveillance (Vrabko et al., 2019). Nowadays, the idea of digital and accessible

Table 3.1 Elements of the state capacity

State capacity	Selected objectives
Administrative capacity	Day-to-day administration of the state
Legal capacity	Rule of law Civil rights and liberties Access to justice Trustworthy court system Functioning prison system
Infrastructural capacity	Highway and railway system Shipping and air transport Telecommunications and post services Digital infrastructure Critical infrastructure
Fiscal capacity	Sufficient fiscal revenues Avoiding excessive external dependence Acceptable socioeconomic inequalities
Military capacity	Defensible army Trusted and accountable police Deployment and resilient units

Source: Own elaboration

administration of the state is especially prevalent. Self-evidently, the administrative capacity of the state is not function only of quantity but equally importantly also of quality, as the bureaucratic apparatus lacking human capital would not be able to properly decode all the intentions of the state and implement them accordingly (Weber, 1968; Rauch & Evans, 2000).

Second, the legal capacity is a necessary condition for securing the rule of law in the country. The legal capacity, similar to the administrative capacity, covers a wide range of state functions, such as the capacity to enforce contracts and property rights, the ability to maintain a trustworthy judicial system for settling disputes as a prerequisite of the rule of law, as well as the strength to ensure that the law applies to everyone equally, which is a necessary condition for free and just society. Apart from self-evidently important legal functions, such as the protection of fundamental rights and liberties, the state capacity is relied on to deliver much more nowadays regarding the legal capacity, for instance to make sure that every citizen is able to access justice regardless of their financial means. Among other things, the legal capacity also entails the capacity to ensure the laws are comprehensible to everyone, the need to secure access to legal aid in civil, criminal, and administrative matters, the right to a decent living, as well as properly functioning and reliable prison system (Jüriloo, 2015; Collier, 2009).

Third, infrastructural capacity represents, according to Soifer (2008), the territorial reach of the state, ipso facto the extent to which control can be exercised over the territory. In its traditional meaning, the infrastructural capacity encompasses the total reach of highways and railways on its territory, the shipping and air transport, as well as the coverage of the telecommunication and post services. It is only natural that the European state in the twenty-first century is expected to deliver much more with regard to its infrastructural capacity, namely the digital infrastructure (EC, 2022b). The digital infrastructure is necessary to enhance the digital skills of the EU's citizens in order to lower the discrepancies in the job markets (OECD, 2022a) and also to secure access to the administrative capacities of the state. In addition to the aforementioned infrastructural capacities (transport and digital), the so-called critical infrastructure in general is necessary to be secured properly as part of the infrastructural capacity. Critical infrastructure is an all-encompassing term that goes beyond simplistic partial infrastructures and refers to an intertwined system that was constructed to ensure the safety and economic and social welfare of modern society and whose resilience is of national interest to every state. The critical infrastructure usually refers to a wide range of buildings and institutions, from hospitals to power plants, and often includes digital and transport capacities (Labaka et al., 2016).

Fourth, fiscal capacity is generally considered to be the state's ability to raise revenues from taxes and subsequently use the collected fiscal resources according to the predefined goals of the government. Without the fiscal capacity to raise revenues or to borrow money from internal and external lenders, the state could not secure all the other aspects of its capacity – neither to pay out its bureaucratic apparatus nor to construct its infrastructure (Besley & Persson, 2011). Of course, not all the resources need to be raised through taxes, and some sources of finance can be extracted from uneven power of balance within a monetary union (e.g., Matthijs & Blyth, 2015), attraction of finance through low corporate taxes, extraction of raw materials or external financing from either the remittances, foreign direct investment, or development aid. In a nutshell, it is important for the state to have sufficient fiscal capacity to implement its goals, sustain socioeconomic cohesion and acceptable levels of socioeconomic inequalities that will not be contested by its citizens, and at the same time be cautious regarding the source of its fiscal revenues. For instance, excessive dependence on external sources of finance (e.g., foreign direct investment) can put a constrain on the state capacity to use these resources for the public good and reinforce its resilience as well as to collect additional finance in the future due to over-indebtedness (Nölke & Vliegenthart, 2009).

Lastly, we also need to acknowledge the importance of military capacity as the state's ability to defend its territory and citizens from external as well as internal threats. Naturally, the army is the most visible component of this capacity when it comes to the defense against foreign intruders and unexpected internal disasters. In certain cases of smaller European countries, the national defense does not have to be secured by the state military units directly (not entirely) but can be quasi-outsourced to other transnational non-state actors, such as the North Atlantic Treaty Organization (NATO). The military capacity in our understanding, however, covers all the law-enforcing elements of the state, not only the army per se but also the police departments (e.g., financial police, customs police, counterterrorism units, investigators, and operative units), intelligence services, rapid deployment units, or firefighter departments (Hendrix, 2011). In a nutshell, all the law-enforcement units working under the umbrella of the state are responsible for keeping the territory of the state and its citizens crime-free and safe. Needless to say, the military capacity is a very important aspect in resilience building vis-à-vis social tipping points given their abrupt and nonlinear development. The rapid deployment forces, police, army, and firefighter departments and their partial capacities are absolutely necessary for the state to be resilient against such unexpected threats for its citizens.

3.2.2 *State Capacity and Democracy*

In the context of the erosion of democracy in recent years, we consider it to be also important to briefly accentuate the link between state capacity and democracy. When we take a look around the globe, we can immediately notice there is a relatively significant correlation between the state capacity and the quality of democracy. To put it bluntly, there are very few autocracies with effective institutions, while most authoritarian regimes lack the capacity to implement their authority. Even authors such as Bäck and Hadenius (2008) and Møller and Skaaning (2011) argue that high levels of state capacity and quality of institutions are necessary prerequisites for democracy and vice versa. Democracy is built on the functioning state capacity. There are certain exceptions of "successful" autocratic regimes in this regard though (e.g., Bueno de Mesquita et al., 2003; Egorov et al., 2009; Mihályi & Szelényi, 2020), however, this monograph mostly deals with governance and resilience in European countries, where democratic regimes still prevail and the capacity to implement governance or the authority of the state is (almost) always followed by the feedback and accountability. Hence, we must be aware that democratic accountability and economic structure tend to go hand in hand (Przeworski et al., 2000). That is why advanced (post)industrial countries in Europe tend to, in general, exhibit high(er) state capacity in the implementation of their policies, at least compared with the less economically developed countries (Meckling & Najm, 2018). It can be, therefore, argued that democratic and economically (more) developed countries are, on average, more successful in implementing their states' authority, due to higher state capacities in all its multifaceted aspects (i.e., administrative, legal, infrastructural, fiscal, and military). Thus, the emphasis on democratic processes and accountability is strongly present in our theoretical and empirical assessment of governance and the state capacity throughout this monograph.

3.3 State Capacity Absence and State Fragility

It should be self-evident from the previous few paragraphs that the concept of state capacity is multifaceted and therefore dependent on countless parameters, regardless of the type of regime as each regime may have a very different approach toward governance. When it comes to major determinants of state capacity, it mostly depends on the social, political, and economic conditions of the country, the state of the elite, or the number of skillful professionals working for the state. These conditions are crucial in boosting the state's capacity (Luiz,

2000). However, there always comes a time when the state capacity of some states deteriorates, and the state is not able to implement its policies as part of its governance. The lack of state capacity is an emblematic feature in the so-called failed states (Niemann, 2007) incapable of performing their functions and delivering public goods (Ignatieff, 2004).

According to Rotberg (2004), there are various stages of state failure, differing in levels of their effective delivery of the most crucial political goods (e.g., security, political and civic freedoms, health care, education, infrastructure, functioning banking system, and fiscal institutions). The author subsequently identifies strong, weak, fragile, and failed or collapsed states, based on their ability to deliver the aforementioned political goods. Strong states, according to Rotberg (2004), perform well across these categories. Weak states show a mixed profile, fulfilling expectations in some areas and performing poorly in others, while the more poorly a weak state performs, the more that weakness tends to tilt toward failure and collapse. Weak states can be inherently weak because of geographical, physical, or economic constraints, or they can be fundamentally strong, but temporarily weak because of internal antagonism, management flaws, greed, despotism, or external attacks. Weak states tend to be poorer, unable to maintain certain socioeconomic functions, and raise revenues to deliver certain needs to their citizens (Besley & Persson, 2010). A version of the weak state with even a "weaker" state capacity is a so-called fragile state. The World Bank (WB) defines a fragile state as one "facing particularly severe development challenges, such as institutional capacity, poor governance, and political instability." WB includes in its list of fragile states countries with high levels of institutional and social fragility and countries affected by violent conflicts (Mackinder, 2020). The International Monetary Fund (IMF) adopts a very similar approach to that of the World Bank and argues that fragile states have characteristics that substantially impair their economic and social performance, including weak governance, limited administrative capacity, chronic humanitarian crisis, persistent social tensions, and, often, violence or the legacy of armed conflict and civil war. The European Parliament (EP), on the contrary, characterizes fragile states as weak or failing structures where the social contract is broken due to the state's incapacity to deal with its basic functions, meet its obligations and responsibilities regarding service delivery, management of resources, rule of law, equitable access to power, security and safety of the populace, and protection and promotion of citizens' rights and freedoms (EP, 2013). The Organization for Economic Co-operation and Development OECD (2022b) characterizes state fragility as the "combination of exposure to risk and insufficient coping capacity of the state, systems or communities to manage, absorb or mitigate those risks." Fragility can ultimately lead to negative outcomes including violence, poverty, inequality, displacement, and environmental degradation. Failed states are a "special category of states that are tense, deeply conflicted, dangerous, and contested by warring factions" (Rotberg, 2004). For the most part, these countries are unable to control their peripheral regions, especially those regions occupied by out-groups, and tend to lose authority over large sections of the territory, so their nominal borders become irrelevant. Another indicator of state failure is the occurrence of criminal violence,

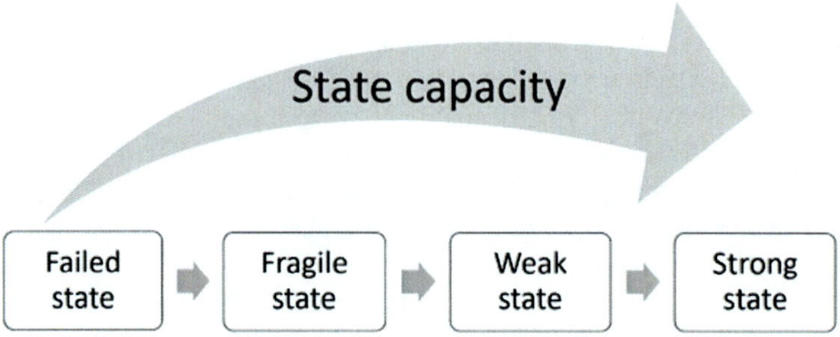

Source: Own elaboration.

Fig. 3.2 State's strength as a function of state capacity. (*Source*: Own elaboration)

murders, gun violence, thefts, or rapes. As state authority weakens and fails, lawlessness becomes more and more apparent. For protection, citizens frequently turn to privateers, warlords, and other local actors based on ethnic or clan solidarity, thus supplementing the possibility of security at a time when the state is on the verge of collapse. The ultimate feature of a state failure is the loss of legitimacy. Ultimately, a collapsed state is a rare and extreme version of a failed state (Grimm & Schneider, 2011).

The continuum from a strong state to a failed state is by definition a function of the state capacity (Fig. 3.2). The less a country is capable of delivering desired political goods to its polity, ergo being unable to implement its will over its territory and population, the more likely the state becomes weak, fragile, or ultimately failed (see also Szabó & Jančovič, 2020). Naturally, in the context of our countries of interest, countries within the EU, it would not be appropriate to talk about failed or even fragile and weak states. In spite of many flaws, the EU member states are among the most developed and politically liberalized state entities in the world. Our argument, however, is that any European state can under the right combination of internal and external risks and abrupt changes become unable to sufficiently manage, absorb, or mitigate impacts stemming from social tipping points and move downward on the continuum toward weakness and fragility. This is where we depart from the "traditional" approaches to the state failure and state fragility (Grimm & Schneider, 2011) and redirect our attention to cases where the state fails to implement *some* of its policies due to a decline in its state capacity being caused by abrupt changes stemming from social tipping points. This, by definition, links the state's weakness in certain policy areas with the state capacity. As the state's capacity (administrative, legal, infrastructural, fiscal, and military) decreases as a result of the state being overwhelmed by the impacts resulting from unanticipated disturbances, its fragility rises, further fueling additional decay of the state capacity.

3.4 State Capacity and Resilience

We know now that the decreased level of the state capacity is correlated with the weakness of the state and by this account also with the state's inability to deliver basic goods to its polity, to implement its defined policies, and, in general, to govern. The state capacity is, therefore, strongly linked to the concept of *resilience*. Albeit one of the most utilized terms in social sciences today (e.g., Martin & Sunley, 2015), there is no universally acknowledged definition of resilience, attracting criticism from various directions (Davidson, 2010; Gong & Hassink, 2017). Resilience is derived from the Latin word *resilire* and refers to leaping back, recovering from, or bouncing back following a shock to its preexisting state (Martin & Sunley, 2015). Generally, therefore, the concept of resilience includes complexity, connectedness, adaptation, or feedback (Brown & Westaway, 2011). The term "resilience" has evolved from the "engineering" definition popularized by Holling (1973: 14), who understands resilience "as a measure of the persistence of systems and of their ability to absorb change and disturbance and still maintain the same relationships between populations or state variables." Basically, how fast a system that has been displaced from equilibrium by a disturbance manages to return to its former equilibrium or the ability of something to get back into its original shape once it has been pulled, stretched, pressed, bent, or otherwise adjusted. Within the engineering approach, it can also be perceived as the capacity of any material to absorb energy when it has been drastically deformed and then, upon release, to have the energy recovered (Taşan-Kok et al., 2012). While this engineering conceptualization of resilience tends to prevail, it is not confined to the physical sciences and has already penetrated multiple fields, including social and environmental sciences.

In (behavioral) psychology, resilience is usually viewed as "the process of, capacity for, or outcomes of a successful adaptation despite challenging or threatening circumstances" and represents the capacity of an individual to maintain or swiftly regain psychopathological well-being following personal trauma (Masten et al., 1990: 426). Resilient individuals demonstrate dynamic self-adjustment, whereas less resilient individuals find themselves negatively impacted by life stressors (Wright et al., 2013). This is what Martin and Sunley (2015) coin as (positive) "adaptive resilience." The conception of resilience as an adaptive response to shocks is also present in complex adaptive systems theory and evolutionary theory. Complex adaptive systems theorists, focusing on the concept of robustness, consider it to be an ability of the system not to be disturbed by external perturbations, therefore, a consistency and stability of its structure. If necessary, robustness requires the ability to undergo plastic changes in some of its features in order to maintain or regain its core performances (Whitacre, 2012). The necessity of evolution and changes is, therefore, crucial to maintaining key functions or performances of a system in complex adaptive system literature (Kitano, 2004). The robustness can be, according to Martin and Sunley (2015), manifested in various ways, namely robust transition (bounce forward) and robust adaptation (bounce back). Thus, the strand of literature concerning complex adaptive systems perceives resilience as the capacity of a system to undergo successful change in structures and functions when faced with perturbations. In the socioeconomic context, this would mean the agency of certain stakeholders, in anticipation of certain types of shocks (migration, climate change, or war), to deliberately undertake changes or a transformation in order to minimize any future shocks of similar magnitudes (e.g., extending capacities of refugee reception centers, enabling local stakeholders to remove fallen trees and lowering dependency on gas supplies from potential geopolitical rival).

Other social sciences, however, have slightly different views regarding resilience. Resilience in economics can be defined as the ability of an economy to retain high employment and low inflation, ergo macroeconomic stability while facing disturbances such as supply and demand shocks (Baggio et al., 2015). Even Holling himself argues that the idea of resilience from the engineering perspective resembles the idea of self-restoring market dynamics in economics, where the primary assumption is that the economy returns to its steady state (equilibrium) in the long run if distorted from equilibrium by shocks (Martin & Sunley, 2015). A similar approach is also in the field of ecology. The idea of resilience found its interpretation in disaster prevention and climate change adaptation (Pelling, 2010) or in the capacity of some ecosystem and social–ecological systems to maintain and subsequently regain stability when faced with exogenous or endogenous disturbances (O'Neill et al., 1986). The works in ecology originating since the 1980s developed the concept of so-called extended ecological resilience (e.g., Holling, 1986; Gunderson & Holling, 2002). According to Walker et al. (2004, 2006), this ecological resilience within the social–ecological systems should be viewed as "the capacity of a system to absorb disturbance and reorganize while undergoing change so as to still retain essentially the same function, structure, identity and feedbacks." Adger's (2000: 361) perception of resilience is similar to that of the ability of communities to withstand shocks to their social infrastructure, recognizing social resilience as an analog to ecological resilience that is linked through closely coupled systems of natural resources and the communities that depend on them. Fjäder (2014: 120), similarly, understands resilience as a "combination of an ability to resist, recover from and reorganize in response to a shock or a crisis." These definitions tend to synthesize the concept of resilience as the amount of disturbance that a system can undergo while retaining its previous functions (Gunderson & Holling, 2002) and in some cases even the extent to which a system is able to reorganize and adapt (Carpenter et al., 2001). Thus, it is obvious that compared with the complex adaptive systems, the extended ecological definition of resilience prioritizes stability and persistence in the face of disturbances, not necessarily a change.

While these three approaches to resilience (pure engineering, adaptive resilience, and extended ecological resilience) might seem mutually exclusive, it does not necessarily have to be the case. As we already explained in the previous chapter, our ambition is to move toward interdisciplinarity, and we thus intend to combine the social–ecological aspects of dynamic systems with the complex adaptive system approach. In fact, the idea of adaptive resilience and robustness in some sense subsumes aspects of the other two interpretations (Martin & Sunley, 2015). The cornerstone of our understanding of systemic resilience, therefore, represents the ability of the system to withstand disturbances stemming from the abrupt nonlinear changes in social–ecological phenomena of the system – social tipping points – that have self-reinforcing feedback mechanism (absorptive capacity). However, we cannot limit ourselves to this narrow idea of resilience as the concrete wall with its absorptive capacity and we also must take into account the adaptive aspect as envisaged by the adaptive system approach. In line with the Commission's strategic foresight report (EC, 2020), we, therefore, also put an emphasis on the ability to transform when faced with perturbation while still retaining key features of the system. These key features are environmentally and digitally sustainable, socially cohesive, strategically autonomous, and democratic systems. Thus, in combination with the absorptive capacity, the adaptive and transformative capacities are also required (e.g., Manca et al., 2017).

By definition, the concept of resilience is dependent on the intensity of a disturbance (social tipping point), as a shock-free world would not require resilience at all. Hence, there is always a certain riskiness of being hit by a disturbance, on the one hand, and the degree to which the system is vulnerable or not toward the disturbance, on the other. Based on the conceptualization developed by Manca et al. (2017), the system is resilient if the combination of riskiness and vulnerability is low or alternatively even when the vulnerability is high, but only limited losses are borne. Depending on the intensity of disturbance, we can distinguish among three levels of exposure to shocks and subsequent levels of required capacities to be resilient – absorptive, adaptive, and transformative (Fig. 3.3). Initially, if a system is

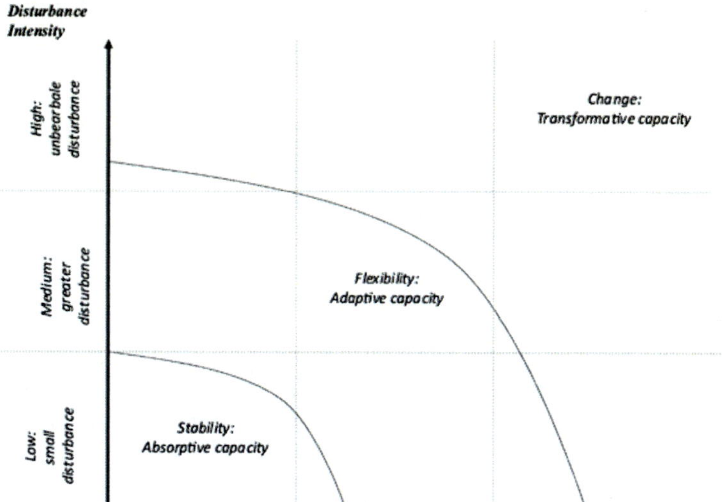

Source: Own elaboration based on Manca et al. (2017: 8).

Fig. 3.3 Conceptualizing capacity, resilience, and disturbances. (*Source*: Own elaboration based on Manca et al. (2017: 8))

faced with a small disturbance for a limited amount of time, the resilient system is able to absorb the shock. As the time of exposure and its intensity rise, the absorptive capacity might be exceeded. Subsequently, if a sizeable disturbance occurs, the resilient system will either adopt or transform. The adaptive capacity refers to a situation in which actors must undergo incremental changes and adjust their expectations according to the worsened situation which requires substantial flexibility on their part so they can mitigate potential damages (Bené et al., 2012). If "the disturbance becomes unbearable (both in terms of its intensity and persistence) and the adaptation would lead to a too large change, a transformation happens" (Manca et al., 2017: 8), which refers to learning from the past events and tailoring transformations given the current constraints imposed by the unbearable disturbance. Such a transformation might be voluntary or enforced by social–ecological conditions. Alternatively, a system with insufficiently large resilience might collapse under the heavy burden of a social tipping point, causing massive social distress. For instance, climate change disturbances of large intensity might lead to the entire collapse of the social component (e.g., local economy) of the system in case of insufficient resilience.

In the previous chapters, it has been argued that state capacity represents the ability of the state to implement its defined goals (Skocpole, 1985), whereas the capacity has a multi-faceted structure (administrative, legal, infrastructural, fiscal, and military). Naturally, similar to our methodological grasp of the concepts of governance and state capacity, we also consider resilience, socially speaking, to be first and foremost a state-centric concept. The state is the primary social entity that is expected and constructed to have the ability and capacity to be resilient in case of some unexpected events (Fjäder, 2014). The state represents the Hobbesian Leviathan – "there is no power on earth compared to him" (Acemoglu & Robinson, 2020: 10). Once abrupt changes emerge, such as a climate-induced social tipping point, the state shall be at the forefront of defending its polity. In this regard, we understand the concept of resilience, in its state-centric form, as the amount of the state's capacity to withstand disturbances (absorptive capacity) while being able to retain its functions and adapt (adaptive capacity) and transform under the changing circumstances (transformative capacity). Understanding the concepts of state capacity and (state) resilience as we do, one has to realize that state capacity and resilience are two sides of the same coin as the state capacity is a prerequisite for the state to be, become, or remain resilient (Haldrup & Rosén, 2013). For instance, without a sufficiently strong and prepared administrative apparatus, the state would become overwhelmed with migration and refugee requests if the migration crisis hits. The same also applies to the legal, infrastructural, fiscal, and military capacity of the state.

3.5 Non-State Actors' Capacities and Resilience

It was shown that the state strength is a function of the state capacity and by that account resilience. However, what if the state's strength is insufficient, causing impediments to its capacity and resilience, which might turn out to be problematic in case of climate, migration, or geopolitics-induced social tipping points? We will argue that this is a place where the non-state actors need to be factored in, since their capacity to implement the state's governance, and to deliver certain public goods otherwise provided by the state as well as their ability to withstand external disturbances and adapt to them might ultimately prevent the state from collapsing under the heavy impacts stemming from social tipping points and accompanying side effects (e.g., David, 2018). "Resilience requires a variety of capacities" (Manca et al., 2017: 9), and by this definition, societal capacity and resilience equal the sum of partial capacities of the state and non-state actors. That is why a multilevel capacity and resilience need to be considered (Ellis et al., 2022), and actors playing the role in the system resilience need to be identified (Manca et al., 2017). The argument that non-state or nonparty actors can enhance the overall societal capacity of governance and resilience is long present in climate change literature (e.g., Bowman & Minas, 2019) and multilateral agreements in which states commit to fight climate change along with other non-state actors, namely within the United Nations Framework Convention for Climate Change (UNFCCC) and subsequent Paris Agreement (UNFCCC, 2018; Hermwille et al., 2019). For instance, the participation of "non-party stakeholders" was cemented in the preamble of the Paris Agreement itself (Phillips, 2018). However, we argue that the joint capacity and resilience of the state and non-state actors can be beneficial and can also potentially enhance governance in other pivotal areas, especially with regard to abrupt nonlinear triggers in the social–ecological systems.

According to Josselin and Wallace (2001: 3), the non-state actors are agents that are "at least in principle autonomous from the structure of the state, and of the governmental and intergovernmental bodies below and above the formally-sovereign state." Some of these actors are transnational – they conduct their activities on a cross-border basis – and some are inwards-oriented or national. Based on this, we can differentiate between the transnational and national non-state actors. Furthermore, in compliance with the UNFCCC conceptualization of the "non-party stakeholders" (Ellis et al., 2022), we can think of the national non-state actors as both civil society-based (businesses, NGOs, academia, and organized youth communities) or subnational (regional, municipal, and local) actors. Furthermore, not all transnational non-state actors are equal, and we also need to factor in the quasi-constitutional role of the EU with regard to its member states (Albi & Bardutzky, 2019). As is shown in Sect. 3.5.2, the EU occupies a special position with regard to its member states given the integration of certain core state powers, making it important to distinguish between the EU as a transnational actor and the remaining transnational actors. As a result, the variety of non-state actors possibly contributing to the state capacity can be narrowed down into two categories – national and transnational – whereas the national can be further subdivided into local and private non-state actors and the transnational into the EU and other transnational non-state actors (Fig. 3.4).

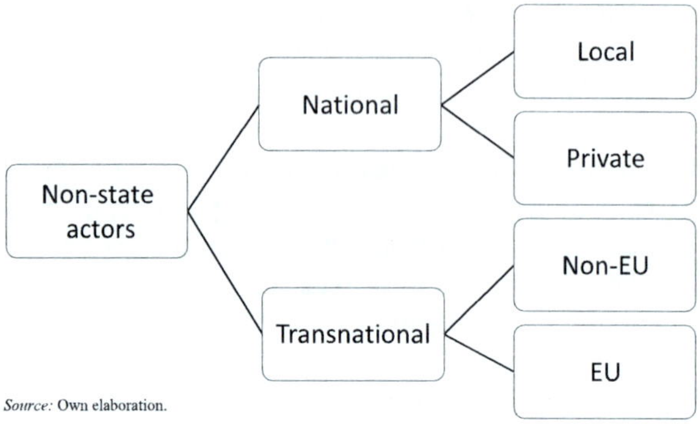

Fig. 3.4 Conceptualizing non-state actors within the EU member states. (*Source*: Own elaboration)

3.5.1 Local Non-State Actors

Non-state actors that are locally present have the most direct access to the citizens and their needs regarding the delivery of public goods. It is not surprising that according to the European Committee of the Regions (CoR), the EU regions were one of the most hit by the Russian aggression in Ukraine in terms of refugees' flows, energy, and food security as well as the economic impact (Kańduła & Przybylska, 2021). From the outset, mayors and regional leaders in the peripheral regions of the EU played a key role in addressing the challenges of providing accommodation, means of subsistence, medical care, education, and employment for the refugees at the EU–Ukraine border. Not only did the peripheral regions in the east of the EU take in most of the refugees, but it is also well documented that they had to absorb the biggest consequences stemming from the economic sanctions imposed on Russia in terms of the GDP drop and inflation spikes (CoR, 2022a). The same applies to the impact of the COVID-19 pandemic on local and regional authorities, who were in the first line of implementing the contagion measures, testing for COVID-19, and then organizing the anti-COVID vaccinations, with similarly malign impact on economic, social, and health dimensions of their citizens' lives (Ibidem). All this was subsequently reflected in the regional and local barometer among local politicians who were asked what level born the greatest burden regarding these two tipping points (CoR, 2022b). Local authorities are usually the ones that directly provide selected public goods and many important aspects of state capacity, in areas such as notary offices, health centers, health posts, schools, libraries, fire stations, jails, tax collection offices, and part of the public bureaucracy. The local regional authorities are also crucial in the green transition. According to the REPowerEU action plan and the accompanying EU Save Energy, besides the member states, the local and regional authorities and organizations are in the best position to reach out to citizens and private sectors to encourage them to undertake energy efficiency measures as they have deep knowledge about people facing risk of energy poverty (EC, 2022a). The authorities in the self-government municipalities are democratically elected and, therefore, in direct and day-to-day contact with their constituency, which makes them more prone to deliver demanded public goods. Thus, proper decentralization and political regions with sufficient capacities are at the center of every well-functioning democracy. According to the European Charter of Local Self-Government which lays down standards for protecting the rights of local authorities and requires the states that have ratified it to comply with its principles and practices, the role of communities at the local level is necessary for the exercise of democracy, and in light of the principle of subsidiarity, it must be allowed for the decentralization of power to the closest to the citizen, and this transfer of competencies must be accompanied with an adequate transfer of financial resources (ECLSG, 1985). Whatever type of political local authorities, be it regions, provinces, countries, or districts (see further Eurostat, 2022), their respective capacities and resilience are crucial for the state and the society as a whole as it is well documented that internal and local disputes erode the state capacity for a very long time (Besley & Persson, 2008), and reinforcing the capacity of local authorities must be a strategic choice for every state wishing to become more resilient (Acemoglu et al., 2015).

In addition to local political authorities, we cannot omit the respective capacities of representatives of local civil society, namely local nonprofit organizations (NPOs), nongovernmental organizations (NGOs), and unorganized local volunteers whose capacity and resilience are also crucial in the case of social tipping points. The importance of local nonpolitical actors with regard to resilience has been demonstrated in numerous types (e.g., de Milliano & Jurriens, 2016). NPOs and NGOs, for instance, have a significant positive relationship with state capacity, especially in democracies, which are countries of our interest (Campbell et al., 2018). The NGOs are mostly invested in human rights protection and the rule of law, the promotion of policies mitigating climate change, or the defense and hybrid warfare, thereby indirectly increasing the capacity and resilience in these areas. As will be shown in Sect. 6.3, in addition to regional local authorities and institutionalized NPOs and NGOs, the inclusion of local volunteers willingly helping the society to uphold its resilience can be decisive in times of crises. Local communities with no formal institutionalization are generally considered to be at the heart of the preparedness for and response to disasters (Comes, 2016).

3.5.2 *Private Non-State Actors*

Besides a number of local non-state actors (regional local authorities, NGOs, and unorganized local civil society), for-profit private actors are also crucial for societal resilience, especially in European countries relying on market redistribution as a primary source of socioeconomic exchanges. Under the label of for-profit private actors, we can subsume corporations, associations of employers, as well as self-employed privateers who create and contribute to private, mostly economic resilience of a country. There are informed opinions that the for-profit private sector might have the capacity to contribute to the achievement of a more resilient society not only in the economic area (May, 2012). Naturally, the primary goal of all private companies is to seek profit and not to invest their own resources in resilience building; however, dexterously constructed public–private partnerships (PPPs) tend to result in increased state capacity and ultimately resilience building with regard to administrative, infrastructural, and even military capacities. It has been shown that the share of privately owned and operated infrastructure has increased globally, and among the OECD member states, the critical infrastructure is mostly owned by the private sector. Fjäder (2014: 124) argues that "whilst the private ownership of infrastructure has arguably created financing and operational efficiency, it has also intensified the state's dependency on private institutions for the delivery of basic services, not only under normal circumstances but also under exceptional circumstances." According to Zhou (2012), however, private non-state actors are not only for-profit but also nonprofit organizations whose members are civil society individuals. At this point, there is, therefore, a necessary overlapping between the local non-state actors and private non-state actors, especially with regard to NPOs and NGOs. As a result, we ought to distinguish between local nonprofit non-state actors and private nonprofit non-state actors. While the former is locally embedded and develops ecological, social, or infrastructural activities that are linked to its local communities, the latter has a slightly wider range of coverage and does not limit itself to a certain region but attaches itself to the civil society in general. Such a categorization can often been viewed as arbitrary should thus be considered as a mere theoretical exercise.

3.5.3 Non-EU Transnational Non-State Actors

Transnational actors comprise numerous private, intergovernmental, and global institutions that have emerged in world politics and are not necessarily congruent with state territorial space (Rosenau, 2003). Among the transnational actors operating within the European context, the European Union is undoubtedly the most impactful one. The EU's impact on the political, social, and environmental development of its peripheral member states is immense and beyond the scope of this book (e.g., Rapacki et al., 2019). It has been shown many times that the EU shapes the domestic political and institutional factors that, in turn, contribute to the enhancement of domestic state capacities in development (Bruzst & Langbein, 2020) or industrial policies (Medve-Bálint & Šćepanović, 2020). The role of the EU in the migration crisis and the migration policies of the EU deeply linked to the states' capacities to cope with the unprecedented flows of refugees and migrants should not be overlooked as well (Geddes et al., 2020). However, as was already hinted above, given the *sui generis* nature of the European integration, the EU needs to be given special attention and is thus conceptualized in a separate section (Sect. 3.5.2).

Apart from the EU, different variants of transitional governance and ipso facto capacities and resilience can be distinguished. First, we need to focus on international organizations (IOs) with the states as their direct member – intergovernmental IOs. There are numerous theories and theoretical frameworks dealing with the issue of IOs that are beyond the scope of this book (e.g., Cogan et al., 2016). We, therefore, limit ourselves to the nuts and bolts of the IOs and their potential capacities. The IOs are formal organizations having a formal secretariat, which were set up by a treaty by at least three members, sovereign states, and by facilitating multilateral action are expected to constrain or regulate the exercise of discretion by those states (Pevehouse & Von Borzyskowski, 2016). However, the IOs are not only about putting constraints on their members; on the contrary, the member states tend to pool their resources into the IOs if their preferences match in order to enhance their respective capacities in selected areas (Moravcsik, 1998). As of 2014, the Union of International Associations (UIA) cataloged 1193 IOs functioning in the world (UIA, 2014). These organizations have activities in almost all important areas of politics, ranging from trade, security, finance, environment, human rights, and science and development, touching upon the state capacity in all of its five areas.

For instance, the activities of the United Nations High Commissioner for Refugees (UNHCR) with its respective administrative are indispensable for the member states in the case of social tipping points concerning a migration situation, given its central role in creating the international refugee regime (Loescher, 2016). Furthermore, direct access to legal expertise and justice is also crucial, and a lack of state capacity to deliver prompt justice might be fatal (justice delayed, justice denied). Fortunately, transnational IOs with complex legal apparatus and legal capacity can avoid state failures when it is insufficient when faced with a social tipping point. In fact, the legal capacity, especially with regard to human rights protection and protection of the principles of the rule of law, is currently dominated at the transnational and European levels. Within the European context, the Council of Europe with the European Convention on Human Rights and the European Court of Human Rights (ECHR) is dominant on the European continent regarding the influence on legal systems, legal expertise, and legal capacity of its member states (Shelton, 2016). The legal influence, either via the case law of the ECHR or recommendations concerning the constitutional matters by the European Commission for Democracy through Law – the Venice Commission – has the capacity to impactfully shape the legal systems and by that account also the legal capacity of its member states. The infrastructural and fiscal capacities of the state are also enhanced by various IOs. International financial institutions, such as the International Monetary Fund (IMF) or the World Bank (WB), contribute to the governance of international finance by supporting numerous long-term as well as medium-to-short-term developmental projects and also helping countries hit by exogenous economic shocks and subsequent balance-of-payment problems. World Trade Organization (WTO), on the contrary, enhances the capacity and resilience of the world trade system by removing barriers and settling disputes between contestant economies (Moeschella, 2016). Furthermore, there are numerous IOs besides the EU that have the potential to enhance the capacity and resilience of its member states in coercive areas, ranging from peace operations, counterterrorism and transnational crime, and non-proliferation and disbarment. Peace operations are especially crucial in the case of military deployments undertaken to achieve humanitarian relief, conflict stabilization, and ceasefire monitoring in affected states. Ultimately, the peace-making IOs are in control of implementing peace agreements and subsequent state building in war-torn areas, which is especially crucial given one of the case studies in this monograph deals with the war-related social tipping point. Signature IOs in the field of peacemaking are, naturally, the United Nations (UN) and the Organization for Security and Co-operation in Europe (OSCE) as well as NATO to a certain degree (Dayal & Morjé Howard, 2016). With its broad range of activities, the UN also plays a pivotal role in global disbarment activities pushing nuclear powers toward the nonproliferation treaties (Shidhu, 2016). IOs such as Interpol help its member states in conducting activities to fight crime within the international space, such as narcotics, money laundering, or terrorism (Cockayne, 2016).

Second, there is a wide diversity of private actors – international nongovernmental organizations (INGOs), multinational corporations (MNCs), and other transnational networks – exercising authority in world politics. This authority can be directly delegated from the state actors or be a result of private initiative. The "delegated private authorities," especially within the globalized economy, tend to take on the role of regulators, implementors, and even the enforcers of global rules, especially in global efforts to tackle climate change as they tend to fulfill the role of expertise throughout treaty negotiations and post-treaty implementation. Experts from INGOs, the private sector, and companies as well as the scientific community are often called upon to provide "expert opinion" to which policymakers listen, thereby reinforcing their respective capacities (Green, 2014). Naturally, the private authorities as non-state actors in world politics are also fueled by the state. As part of the public–private governance partnerships, the state tends to delegate some authority to private non-state actors. As opposed to frequent concerns regarding the retreat of the state, this phenomenon appears to be still relatively rare. For instance, Green (2014) observes that of almost 2000 policy functions within the analyzed multilateral environmental treaties, less than 4% were delegated to private actors and less than 0.1% of all policy functions are the sole purview of a private agent as the total amount of governance is increasing. Yet, (private) "non-state agents fill a small, but growing role" on important areas of world politics (Green, 2018). Besides the delegated private authority, which still plays a relatively minor role in global affairs, there is also the "entrepreneurial private authority," or a situation in which private actors create rules or set standards with no explicit delegation from the state authority, these norms and regulations are utterly decentralized (Green, 2014). In these cases of transnational governance, state actors have no or minimal role in managing the private voluntary regulatory schemes. Accurate examples of this kind of entrepreneurial authority can be, for instance, the International Organization of Standardization (ISO) or the Society for Worldwide Interbank Financial Telecommunication (SWIFT). In this case, we talk about "non-state market driven" governance (Ibidem, p. 116). Similar to the delegated private authority which typically enhances the state capacity with expertise, mainly within infrastructural and administrator capacities, the entrepreneurial private authorities might also support the state in its regulatory capacities, even though it is not under the state control if certain preconditions are met (Verbruggen, 2014).

3.5.4 The Multifaceted Nature of the EU as a Transnational Actor: Integration of Core State Powers

Adapting the rather reductionist state-centric conceptualization of governance and state capacity, we are of the opinion that the states remain to be behind the driving wheel when it comes to enforcing the publicly stated goals. However, as was noted in the previous section that dealt with the transnational non-state actors, the European Union represents a transnational actor *sui generis* as it also assumed some of the "core state powers" via the process of the European integration (Genschel & Jachtenfuchs, 2016) and as the EU legal system enjoys dominance and primacy over the legal system of its member states (Albi & Bardutzky, 2019). In a way, the EU became a form of state of its own – regulatory state (Moravcsik, 2002; Caporaso et al., 2015). Thereby, although we conceptualize the EU as a transnational actor, we consider it to be important to address some of its nuances with regard to the (member) states and the core state powers.

The concept of core state powers is closely intertwined with the state capacity as it revolves around the nationalization of three key policy areas – coercive power, economic and fiscal capacity, as well as the administration might – providing the building blocks for national sovereignty (Krasner, 1999). Coercive powers refer to the state's monopoly of legitimate coercion encompassing military force, police power, and border control, economic and fiscal policies to the currency, monetary policy, and taxation, whereas bureaucratic administration refers to the capacity of implementing and enforcing laws within the national borders (Genschel & Jachtenfuchs, 2014, 2016). This triumvirate thus gives rise to a proper and holistic state. In the beginning, European integration was primarily pursued in economic and commercial areas – market integration. Starting with the Single European Act, however, the European Communities mandated gradually expanded to the integration of core state powers – monetary and fiscal policies, public administration, diplomacy, military and police force as well as border control (Genschel & Jachtenfuchs, 2017).

Starting with the construction of the European Monetary Union (EMU), the EU expanded its capacities to monetary policy areas. In the beginning, European integration in this area was based on asymmetric integration – monetary policy integrated at the supranational EU level (European Central Bank (ECB)), whereas the fiscal capacity remained at national levels (Fabbrini, 2022a, b). This decentralized fiscal status quo was cemented with a set of legally binding fiscal rules designed to ensure that the national budgetary policies are of sound state (Stability and Growth Pact) and the no-bailout rule enshrined in Article 125 of Treaty on the Functioning of the European Union. Although little EU influence was present first over the member states' fiscal capacities, the eurozone crisis (2009–2015) brought about some significant changes in this regard. First of all, major European stakeholders (EU institutions and member states) moved in the direction of strengthening budgetary constraints (Six-Pack, Two-Pack, and Fiscal Compact). Second, temporary and then permanent European emergency funds were introduced – European Financial Stability Facility and European Stability Mechanism – with a lending capacity exceeding half a trillion euros. This was also complemented by a new economic governance mechanism within the EMU, most notably the European Semester, whereby member states are now obliged to submit their draft budgetary plans first to the Commission for review before they are put on the floor in the national assemblies. Third, the integration of financial markets and the creation of the European Banking Union were also major steps on the road toward breaking the sovereign-government nexus. Lastly, all of these steps were backed by unconventional monetary policies of the ECB, which bypassed its treaty mandate and provided fiscal assistance to EMU member states in need (Securities Market Program, Outright Monetary Transaction, and Public Sector Purchase Program). Even the "whatever it takes" forward guidance is a clear indication of how penetrated the core state power in economic governance became vis-à-vis European integration (Fabbrini, 2016). A major shift in fiscal policies, however, occurred following the COVID-19 crisis, when the Commission activated the EU Solidarity Fund in order to provide cash aid to member states, composed a coronavirus response investment initiative (37 billion euros), and established a Temporary Support to mitigate Unemployment Risks in an Emergency (SURE) as an insurance system designed to back up national unemployment insurance regimes that got under the strain due to COVID-related containment measures. These measures were seconded with financial support from the EU financial institutions, namely the European Investment Bank and the ECB with a Pandemic Emergency Purchase Program (Fabbrini, 2022b). Nonetheless, none of these COVID measures constituted a substantial change with regard to fiscal capacity and the core state power. Then, in July 2020, EU heads of state at the European Council stoke a deal on the EU recovery fund worth 750 billion euros (390 billion in grants and 360 billion in loans), Next Generation EU (NGEU), and on the next Multi-Annual Financial Framework 2021–2027 (1074 trillion euros). For the first in European history, the NGEU endowed the EU with something approximating the fiscal capacity, where the Commission is authorized to issue debt on behalf of all the EU and to subsequently raise new taxes aimed at repaying the debt (Fabbrini, 2022a, b). Thus, the Commission is authorized to act as quasi-EU treasury as it can borrow 750 billion euros on behalf of the Union on the capital markets (common debt), use it to run supranational economic policies based on the Commission priorities in green, digital, and just transition, and pay for it with EU resources in the form of a new plastic tax, a carbon border adjustment tax, a digital tax, and potentially a financial transaction tax (Fabbrini, 2022a, b). This way, in the context of the NGEU, the EU economic governance pushed for a rebalancing of the status quo from a governance model based on "unconstrained intergovernmentalism" represented by the ESM toward "constrained supranationalism" under the NGEU (Buti & Fabbrini, 2023).

However, fiscal capacity is not the only core state power that had been penetrated by European integration, but the coercive and administrative powers were also influenced. In spite of being often characterized as ineffective or incoherent, the EU's capacities in military and diplomatic issues are not negligible (Moravcsik & Emmons, 2021). For instance, despite not having a common EU army, the EU has staged numerous overseas military missions since 2003 as part of its peacekeeping operations, conflict preventions, and the strengthening of security as part of the Common Security and Defense Policy operations. Furthermore, within the European External Action Service (EEAS) founded in 2010 (EEAS, 2021), a "quasi-supranational diplomatic corps," challenging the member states' monopoly on foreign representations, was established (Adler-Nissen, 2014). In addition, there has been a significant shift in EU integration in defense after the War on Ukraine (Fiot, 2023). In bureaucratic administration capacities, there is a clear concentration of public administration in the Commission. According to Trondal and Guy Peters (2013), the personnel of the Commission doubled, and the number of EU agencies tripled since the 1990s. Currently, around 32,000 permanent and contract employees work in the Commission as officers, researchers, lawyers, and translators (EC, 2023). Additionally, we cannot omit the emergence of large bureaucracies in the Council, the European Parliament, the ECB, and the Court of Justice of the European Union. Based on older estimations, there are 56,000 people working for various EU institutions, which are further supported by over 60,000 national bureaucrats coordinating the policy implementation between Brussels and member states (Genschel & Jachtenfuchs, 2016). Still, the EU's capacity is relatively insignificant when compared with sovereign states. On the contrary, the EU, as a regulatory state (Caporaso et al., 2015), partly compensates for the lack of material capacity in the core state powers by regulations (Hallerberg, 2014). For instance, codes of the European Defense Agency pressure the member states into extending their military capacities from which the EU draws for joint military missions. The EU also uses its regulatory power to regulate how the member states use their policing powers (Genschel & Jachtenfuchs, 2016).

In light of the conceptualization of the non-state actors, especially those of transnational nature, we considered it to be important to address the *sui generis* nature of the EU as the EU's supranational bodies, namely the Commission, took on numerous competencies that are closely linked to the member states' core state powers. Be it administrative staff that oversees the interdependence between the member states and the EU, legal capacity to pass new laws (namely regulations and directives) in compliance with intergovernmental EU bodies such as the Council, infrastructural capacity to invest through cohesion policies and the Resilience and Recovery Plans, newly emerged fiscal capacity to create a common debt, and subsequent EU sources it finances, as well as military and diplomatic capacity within the EEAS. Speaking of the state capacity of the EU's member states without emphasizing the *sui generis* capacity of the EU would make our analysis incomplete as the boundaries in the conduct of the core state powers and their distribution between the member states' and the EU's competencies get blurrier and blurrier (Genschel & Jachtenfuchs, 2016). Notwithstanding, even though we take note of the *sui generis* nature of the EU that makes it quite distinct from other transnational non-state actors for the reasons explained above (integration of the core state powers), we cannot omit the fact that the member states remain behind the driving wheels of the European integration, thus making the whole concept of governance, capacity, and resilience inherently state-centric first and the EU a mere product of the will of sovereign nations (Moravcsik, 2018). The recent turmoil surrounding Brexit is proof of that. As a result, the EU is considered a transnational non-state actor with *sui generis* position of a supranational legal entity that integrates some of its member state's core state powers.

3.5.5 Inter-dynamics Between State and Non-State Actors' Capacities: A Multilevel Approach

The capacity – the executive branch of governance – spans across multiple levels within the universe of actors representing the European society. At the center, there is the state with the state capacity or the ability to attain the stated objectives of the state which mirrors itself in its resilience. Then, around the state orbit non-state actors located at various levels – transnational or subnational – that are increasingly able to shape the conduct of other actors, but primarily the state (Breslin & Nesadurai, 2018). National actors, both local (regional local authorities, local NGOs, and local civil society) and private (private companies, NPOs, and NGOs), along with transnational actors (IOs and private transnational actors) and the EU have their respective capacities that are reflected in their respective abilities to withstand nonlinear disturbances and adapt. We emphasize the EU among the transnational actors due to its formative impact on the constitutional aspects of its member states.

Using these capacities, the non-state actors at different levels can contribute to the state capacity and resilience and by that account enhance the overall societal capacity and resilience (Fig. 3.5). The sum of all actors' capacities represents the societal capacity to carry out certain tasks, to govern, and, most importantly, to be resilient against various disasters and social tipping points with unintended consequences. Naturally, the type and degree of capacities differ across the non-state actors. It is only natural some transnational actors have more pronounced military capacity (e.g., NATO) than private companies, some regional local authorities build up more robust administrative capacities compared with the EU, and some private nonprofit actors accumulate higher legal capacity than selected MNCs. What is important though is whether the transmission between the state and non-state actors functions properly – the non-state actors are encouraged and willing to contribute to the state capacity – and whether the partial capacities and resilience of the state and non-state actors combined are sufficient to face the disturbances stemming from social tipping points regardless of their origins, be they having origin in migration, climate change or war.

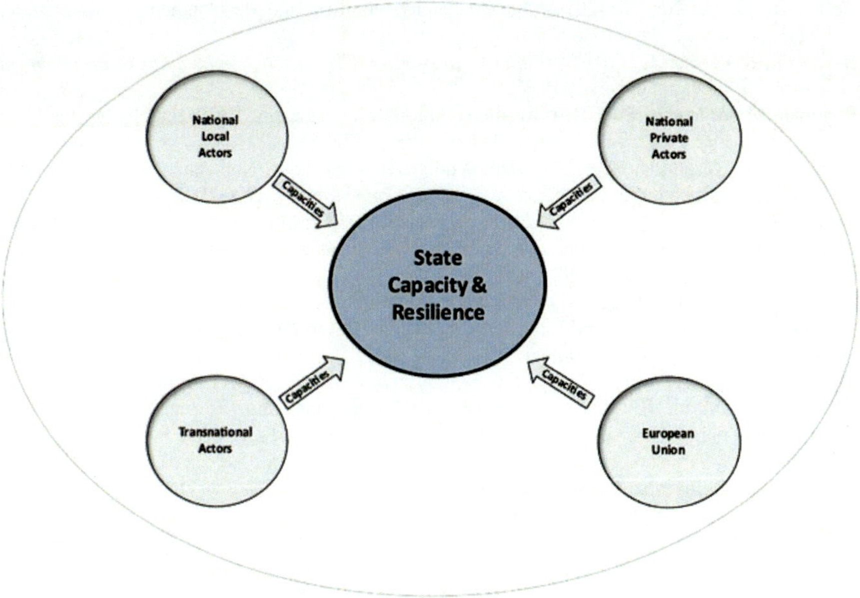

Source: Own elaboration.

Fig. 3.5 Societal capacity and resilience in EU member states. (*Source*: Own elaboration)

References

Acemoglu, D., & Robinson, J. A. (2020). *The narrow corridor: States, societies, and the fate of liberty* (2nd ed.). Penguin Press.

Acemoglu, D., García-Jimeno, C., & Robinson, J. A. (2015). State capacity and economic development: A network approach. *American Economic Review, 105*(8), 2364–2409.

Adger, W. N. (2000). Social and ecological resilience: Are they related? *Progress in Human Geography, 24*(3), 347–364.

Adler-Nissen, R. (2014). Symbolic power in European diplomacy: The struggle between national foreign services and the EU's external action service. *Review of International Studies, 40*(4), 657–681.

Albi, A., & Bardutzky, S. (2019). *National constitutions in European and global governance: Democracy, rights, the rule of law*. Springer Open.

Bäck, H., & Hadenius, A. (2008). Democracy and state capacity: Exploring a J-shaped relationship. *Governance: An International Journal of Policy, Administration, and Institutions, 2*(1), 1–24.

Baggio, J. A., Brown, K., & Hellebrandt, D. (2015). Boundary object or bridging concept? A citation network analysis of resilience. *Source of Society, 20*(2), 2.

Bené, C., et al. (2012). Resilience: New utopia or new tyranny? Reflection about the potentials and limits of the concept of resilience in relation to vulnerability reduction Programmes. *IDS Working Papers, 2012*(405), 1–61.

Besley, T., & Persson, T. (2008). Wars and state capacity. *Journal of the European Economic Association, 6*(2/3), 522–530.

Besley, T., & Persson, T. (2010). State capacity, conflict, and development. *Econometrica, 78*(1), 1–34.

Besley, T., & Persson, T. (2011). *Pillars of prosperity: The political economics of development clusters*. Princeton University Press.

Bowman, M., & Minas, S. (2019). Resilience through interlinkage: The green climate fund and climate finance governance. *Climate Policy, 19*(3), 342–353.

Breslin, S., & Nesadurai, H. E. S. (2018). Who governs and how? Non-state actors and transnational governance in Southeast Asia. *Journal of Contemporary Asia, 48*(2), 187–203.

Brown, K., & Westaway, E. (2011). Agency, capacity, and resilience to environmental change: Lessons from human development, well-being, and disasters. *Annual Review of Environment and Resources, 36*(1), 321–342.

Bruzst, L., & Langbein, J. (2020). Manufacturing development: How transnational market integration shapes opportunities and capacities for development in Europe's three peripheries. *Review of International Political Economy, 27*(5), 996–1019.

Bueno De Mesquita, B., et al. (2003). *The logic of political survival*. MIT Press.

Buti, M., & Fabbrini, S. (2023). Next generation EU and the future of economic governance: Towards a paradigm change or just a big one-off? *Journal of European Public Policy, 3*(4), 676–695.

Campbell, S., Digiuseppe, M., & Murdie, A. (2018). International development NGOs and bureaucratic capacity: Facilitator or destroyer? *Political Research Quarterly, 72*(1), 3–18.

Caporaso, J. A., et al. (2015). Still a regulatory state? The European Union and the financial crisis. *Journal of European Public Policy, 22*(7), 889–907.

Carpenter, S., et al. (2001). From metaphor to measurement: Resilience of what to what? *Ecosystems, 4*(8), 765–781.

Cockayne, J. (2016). Conterterrorism and transnational crime. In J. K. Cogan, I. Hurd, & I. Johnstone (Eds.), *The Oxford handbook of international organizations* (pp. 211–231). Oxford University Press.

Cogan, J. K., Hurd, I., & Johnstone, I. (2016). *The Oxford handbook of international organizations*. Oxford University Press.

Collier, P. (2009). The political economy of state failure. *Oxford Review of Economic Policy, 25*(2), 219–240.

Comes, T. (2016). Designing for networked community resilience. *Procedia Engineering, 159*, 6–11.

CoR. (2022a). EU annual report on the state of regions and cities 2022. *European Committee of the Regions*. Available at https://cor.europa.eu/en/our-work/Pages/State-of-Regions-and-Cities-2022.aspx.

CoR. (2022b). Regional and local barometer. *European Committee of the Regions*. Available at https://cor.europa.eu/en/our-work/Pages/EURegionalBarometer-Survey-2022.aspx.

David, L. (2018). Agency and resilience in the time of regional economic crisis. *European Planning Studies, 26*(5), 1041–1059.

Davidson, D. J. (2010). The applicability of the concept of resilience to social systems: Some sources of optimism and nagging doubts. *Society and Natural Resources, 23*(12), 1135–1149.

Dayal, A., & Morjé Howard, L. (2016). Peace operations. In J. K. Cogan, I. Hurd, & I. Johnstone (Eds.), *The Oxford handbook of international organizations* (pp. 191–210). Oxford University Press.

De Milliano, C. W. J., & Jurriens, J. (2016). Realities of resilience in practice: Lessons learnt through a pilot EU aid volunteer initiative. *International Policies, Practices and Discourses, 4*(2), 79–94.

Dusza, K. (1989). Max Weber's conception of the state. *International Journal of Politics, Culture, and Society, 3*(1), 71–105.

EC. (2020). 2020 strategic foresight report: Charting the course towards a more resilience Europe. *European Commission*. Available at https://commission.europa.eu/strategy-and-policy/strategic-planning/strategic-foresight/2020-strategic-foresight-report_en#documents.

EC. (2022a). EU 'save energy', communication from the commission to the European Parliament, the Council, the European Economic and Social Committee and the Committee of the Regions. *European Commission*. Available at https://eur-lex.europa.eu/legal-content/EN/TXT/PDF/?uri=CELEX:52022DC0240&from=ES.

EC. (2022b). Establishing a European declaration on digital rights and principles for the digital decade, communication from the Commission to the European Parliament, the Council, the European Economic and Social Committee and the Committee of the Regions. *European Commission*. Available at https://digital-strategy.ec.europa.eu/en/library/declaration-european-digital-rights-and-principles.

EC. (2023). European Commission HR key figures 2023. *European Commission*. Available at https://commission.europa.eu/document/download/04118600-5b22-4b63-83e4-bdf74a6be3fe_en?filename=HR-Key-Figures-2023-fr_en.pdf.

ECLSG. (1985). European charter of local self-government (ETS NO. 122). *Congress of Local and Regional Authorities, Council of Europe*. Available at https://www.coe.int/en/web/conventions/full-list?module=treaty-detail&treatynum=122.

EEAS. (2021). Annual activity reports. *European External Action Service*. Available at https://www.eeas.europa.eu/sites/default/files/documents/EEAS%20Annual%20Activity%20Report%202021.pdf.

Egorov, G., Guriev, S., & Sonin, K. (2009). Why resource-poor dictators allow freer media: A theory and evidence from panel data. *American Political Science Review, 103*(4), 645–668.

Ellis, J., Luca, L. R., & De Lorenzo, F. (2022). How national governments can facilitate increased mitigation action from non-party stakeholders: Insights from urban renewable electricity and REDD+. *OECD/IEA Climate Change Expert Group Papers*, No. 2022/02, OECD Publishing, Paris. Available at https://doi.org/10.1787/d2374fc8-en

Enderlein, H., Wälti, S., & Zürn, M. (2010). *Handbook on multi-level governance*. Edward Elgar Publishing.

EP. (2013). EU development cooperation in fragile states: Challenges and opportunities. *Directorate-General for External Policies of the Union*. Available at https://www.europarl.europa.eu/RegData/etudes/etudes/join/2013/433724/EXPO-DEVE_ET(2013)433724_EN.pdf.

Eurostat. (2022). Statistical regions in the European Union and partner countries: NUTS and statistical regions 2021. *Eurostat*. Available at https://ec.europa.eu/eurostat/documents/3859598/15193590/KS-GQ-22-010-EN-N.pdf/82e738dc-fe63-6594-8b2c-1b131ab3f877?t=1666687530717.

Fabbrini, F. (2016). *Economic governance in Europe*. Oxford University Press.

Fabbrini, F. (2022a). *EU fiscal capacity*. Oxford University Press.

Fabbrini, F. (2022b). The legal architecture of the economic responses to Covid-19: EMU beyond the pandemic. *JCMS: Journal of Common Market Studies, 60*(1), 186–203.

Fiot, D. (2023). In every crisis an opportunity? European Union integration in defence and the war on Ukraine. *Journal of European Integration, 45*(3), 447–462.

Fjäder, C. (2014). The nation-state, national security and resilience in the age of globalization. *Resilience, 2*(2), 114–129.

Geddes, A., Hadj Abdou, L., & Brumat, L. (2020). *Migration and mobility in the European Union*. Red Globe Press.

Genschel, P., & Jachtenfuchs, M. (2014). *Beyond the regulatory polity? The European integration of core state powers*. Oxford University Press.

Genschel, P., & Jachtenfuchs, M. (2016). More integration, less federation: The European integration of core state powers. *Journal of European Public Policy, 23*(1), 42–59.

Genschel, P., & Jachtenfuchs, M. (2017). From market integration to core state powers: The Eurozone crisis, the refugee crisis and integration theory. *JCMS: Journal of Common Market Studies, 56*(1), 178–196.

Gong, H., & Hassink, R. (2017). Regional resilience: The critique revisited. In N. Williams & T. Vorley (Eds.), *Creating resilient economies: Entrepreneurship, growth and development in uncertain times* (pp. 206–216). Edward Elgar Publishing.

Green, J. F. (2014). *Rethinking private authority: Agents and entrepreneurs in global environmental governance*. Princeton University Press.

Green, J. F. (2018). Transnational delegation in global environmental governance: When do non-state actors govern? *Regulation & Governance, 12*(2), 263–276.

Grimm, S., & Schneider, G. (2011). Predicting social tipping points: Current research and the way forward. *German Development Institute*, Discussion paper 8/2011.

Gunderson, L. H., & Holling, C. S. (Eds.). (2002). *Panarchy: Understanding transformation in human and natural systems*. Island Press.

Haldrup, S. V., & Rosén, F. (2013). Developing resilience: A retreat from grand planning. *Resilience, 1*(2), 130–145.

Hallerberg, M. (2014). Why is there fiscal capacity but little regulation in the US, but regulation and little fiscal capacity in Europe? The global financial crisis as a test case. In P. Genschel & M. Jachtenfuchs (Eds.), *Beyond the regulatory policy? The European integration of core state powers*. Oxford University Press.

Hendrix, C. S. (2011). Measuring state capacity: Theoretical and empirical implications for the study of civil conflict. *Journal of Peace Research, 47*(3), 273–285.

Héritier, A., & Rhodes, M. (2011). *New modes of governance in Europe: Governing in the shadow of hierarchy*. Palgrave Macmillan.

Hermwille, L., et al. (2019). Catalyzing mitigation ambition under the Paris Agreement: Elements for an effective global stocktake. *Climate Policy, 19*(8), 988–1001.

Holling, C. S. (1973). Resilience and stability of ecological systems. *Annual Review of Ecology and Semantics, 4*, 1–23.

Holling, C. S. (1986). The resilience of terrestrial ecosystems; local surprise and global change. In W. C. Clark & R. E. Munn (Eds.), *Sustainable development of the biosphere* (pp. 292–317). Cambridge University Press.

Ignatieff, M. (2004). State failure and nation-building. In J. L. Holzgrefe & R. O. Keohane (Eds.), *Humanitarian intervention: Ethical, legal, and political dilemmas* (pp. 299–321). Cambridge University Press.

Josselin, D., & Wallace, W. (2001). Non-state actors in world politics: A framework. In D. Josselin & W. Wallace (Eds.), *Non-state actors in world politics* (pp. 1–20).

Jüriloo, K. (2015). Free legal-aid – A human right. *Nordic Journal of Human Rights, 33*(3), 203–219.

Kańduła, S., & Przybylska, J. (2021). Financial instruments used by Polish municipalities in response to the first wave of COVID-19. *Public Organization Review, 21*(4), 665–686.

Kaufmann, D., Aart, K., & Zoido-Lobatón, P. (1999). Aggregating governance indicators. *World Bank policy research working paper*, No. 2195.

Kelemen, R. D. (2016). On the unsustainability of constitutional pluralism: European supremacy and the survival of the Eurozone. *Maastricht Journal of European and Comparative Law, 23*(1), 136–150.

Kelsen, H. (2008). *Pure theory of law*. The Lawbook Exchange.

Kitano, H. (2004). Biological robustness. *Nature Reviews Genetics, 5*, 826–837.

Krasner, S. D. (1999). *Sovereignty: Organized hypocrisy*. Princeton University Press.

Labaka, L., Hernantes, J., & Sarriegi, J. M. (2016). A holistic framework for building critical infrastructure resilience. *Technological Forecasting and Social Change, 103*, 21–33.

Levi-Faur, D. (2012). From "big government" to "big governance". In D. Levi-Faur (Ed.), *The Oxford handbook of governance* (pp. 3–18). Oxford University Press.

Loescher, G. (2016). Refugees and internally displaced persons. In J. K. Cogan, I. Hurd, & I. Johnstone (Eds.), *The Oxford handbook of international organizations* (pp. 323–346). Oxford University Press.

Luiz, J. M. (2000). The politics of state, society and economy. *International Journal of Social Economics, 27*(3), 227–243.

Lynn, L. E. (2012). The many faces of governance: Adaptation? Transformation? Both? Neither? In D. Levi-Faur (Ed.), *The Oxford handbook of governance* (pp. 49–64). Oxford University Press.

Mackinder, S. (2020). Tracing the institutional barriers to the integration of the social protection and fragile states agendas within the World Bank. *Journal of International and Comparative Social Policy, 36*(2), 111–124.

Manca, A. R., Benczur, P., & Giovannini, E. (2017). Building a scientific narrative towards a more resilient EU Society. *European Commission,* JCR Science for Policy Report. Available at https://publications.jrc.ec.europa.eu/repository/bitstream/JRC106265/jrc106265_100417_resilience_scienceforpolicyreport.pdf.

Martin, R., & Sunley, P. (2015). On the notion of regional economic resilience: Conceptualization and explanation. *Journal of Economic Geography, 15*(1), 1–42.

Masten, A. S., Best, K. M., & Garmezy, N. (1990). Resilience and development: Contributions from the study of children who overcome adversity. *Development and Psychopathology, 2*(4), 425–444.

Matthews, F. (2012). Governance and state capacity. In D. Levi-Faur (Ed.), *The Oxford handbook of governance* (pp. 281–293). Oxford University Press.

Matthijs, M., & Blyth, M. (2015). *The future of the euro*. Oxford University Press.

May, P. J. (2012). Public risks and disaster resilience: Rethinking public and private sector roles. In N. Kapucu, C. V. Hawkins, & F. R. Rivera (Eds.), *Disaster resiliency*. Routledge.

Meckling, J., & Najm, J. (2018). The power of process: State capacity and climate policy. *Governance, 31*(4), 741–757.

Medve-Bálint, G., & Šćepanović, V. (2020). EU funds, state capacity and the development of transnational industrial policies in Europe's eastern periphery. *Review of International Political Economy, 27*(5), 1063–1082.

Michalski, W., Miller, R., & Stevens, B. (2001). Governance in the 21st century: Power in the global knowledge economy and society. In *Governance in the 21st century* (pp. 7–26). OECD.

Mihályi, P., & Szelényi, I. (2020). The two forms of modern capitalism: Liberal and illiberal states: A criticism of the varieties of capitalism paradigm. *Comparative Sociology, 19*(2), 155–175.

Moeschella, M. (2016). International finance. In J. K. Cogan, I. Hurd, & I. Johnstone (Eds.), *The Oxford handbook of international organizations* (pp. 365–383). Oxford University Press.

Møller, J., & Skaaning, S. E. (2011). Stateness first? *Democratization, 18*(1), 1–24.

Moravcsik, A. (1998). *The choice for Europe: Social purpose and state power from Messina to Maastricht*. Cornell University Press.

Moravcsik, A. (2002). Reassessing legitimacy in the European Union. *JCMS: Journal of Common Market Studies, 43*(4), 603–624.

Moravcsik, A. (2018). Preferences, power and institutions in 21st – Century Europe. *JCMS: Journal of Common Market Studies, 56*(7), 1648–1674.

Moravcsik, A., & Emmons, C. (2021). Liberal Intergovernmentalism and EU external action. In S. Gstöhl & S. Schunz (Eds.), *The external action of the European Union: Concepts, approaches, theories* (pp. 181–196). Bloomsbury Publishing.

Niemann, M. (2007). War making and state making in Central Africa. *Africa Today, 53*(3), 21–39.

Nölke, A., & Vliegenthart, A. (2009). Enlarging the varieties of capitalism: The emergence of dependent market economies in east Central Europe. *World Politics, 61*(4), 670–702.

O'Neill, et al. (1986). *A hierarchical concept of ecosystems*. Princeton University Press.

OECD. (2022a). Skills for the digital transition: Assessing recent trends using big data. *OECD.* Available at https://www.oecd.org/employment/skills-for-the-digital-transition-38c36777-en.htm.

OECD. (2022b). States of fragility 2022. *OECD*. Available at https://www.oecd.org/publications/states-of-fragility-fa5a6770-en.htm.

Offe, C. (2009). Governance: An "empty signifier". *Constellations, 16*(4), 550–562.

Pelling, M. (2010). *Adaptation to climate change: From resilience to transformation*. Routledge.

Peters, B. G. (2012). Governance as political theory. In D. Levi-Faur (Ed.), *The Oxford handbook of governance* (pp. 17–37). Oxford University Press.

Peters, B. G., & Pierre, J. (2006). Governance, accountability and democratic legitimacy. In A. Benz & Y. Papadopoulos (Eds.), *Governance and democracy: Comparing national, European and international experiences* (pp. 29–43). Routledge.

Pevehouse, J., & Von Borzyskowski, I. (2016). International organizations in world politics. In J. K. Cogan, I. Hurd, & I. Johnstone (Eds.), *The Oxford handbook of international organizations* (pp. 3–32). Oxford University Press.

Phillips, F. K. (2018). Participation of non-party stakeholders under the UNFCCC: Options for future engagement. *Centre for International Governance Innovation*, CIGI Paper No. 205. Available at https://www.cigionline.org/publications/participation-non-party-stakeholders-under-unfccc-options-future-engagement/.

Pierre, J., & Peters, B. G. (2020). *Governance, politics and the state*. Red Globe Press.

Przeworski, A., et al. (2000). *Democracy and development: Political institutions and well-being in the world, 1950–1990*. Cambridge University Press.

Rapacki, R., et al. (2019). *Diversity of patchwork capitalism in central and eastern Europe*. Routledge.

Rauch, J. E., & Evans, P. (2000). Bureaucratic structure and bureaucratic performance in less developed countries. *Journal of Public Economics, 75*(1), 49–71.

Rhodes, R. A. W. (1994). The hollowing out of the state: The changing nature of the public service in Britain. *The Political Quarterly, 65*(2), 138–151.

Rhodes, R. A. W. (1997). *Understanding governance: Policy networks, governance, reflexivity and accountability*. Open University.

Rosenau, J. (2003). *Distinct proximities: Dynamics beyond globalisation*. Princeton University Press.

Rotberg, I. R. (2004). The failure and collapse of nation-states: Breakdown, prevention, and repair. In I. R. Rotberg (Ed.), *When states fail* (pp. 1–50).

Savoia, A., & Sen, K. (2012). Measurement and evolution of state capacity: Exploring a lesser known aspect of governance. *Effective states and inclusive development research centre working paper no. 10*.

Shelton, D. (2016). Human rights. In J. K. Cogan, I. Hurd, & I. Johnstone (Eds.), *The Oxford handbook of international organizations* (pp. 249–281). Oxford University Press.

Shidhu, W. P. S. (2016). Non-proliferation and disarmament. In J. K. Cogan, I. Hurd, & I. Johnstone (Eds.), *The Oxford handbook of international organizations* (pp. 232–248). Oxford University Press.

Skocpole, T. (1985). *Bringing the state back in*. Cambridge University Press.

Soifer, H. (2008). State infrastructural power: Approaches to conceptualization and measurement. *Studies in Comparative International Development, 43*, 231–251.

Stigler, G. J., & Samuelson, P. A. (1968). A dialogue on the proper economic role of the state. In *Selected papers no. 7* (3rd ed.). Graduate School of Business, University of Chicago.

Stiglitz, J., et al. (1989). *The economic role of the state*. Blackwell.

Szabó, J., & Jančovič, P. (2020). Social tipping points & asylum seekers: Fragility in the EU's remote island areas. In *International conference on European integration 2022*, Ostrava, pp. 694–707.

Taşan-Kok, T., Stead, D., & Lu, P. (2012). Conceptual overview of resilience: History and context. In A. Eraydin & T. Taşan-Kok (Eds.), *Resilience thinking in urban planning* (pp. 39–51). Springer.

Trondal, J., & Guy Peters, B. (2013). The rise of European administrative space: Lessons learned. *Journal of European Public Policy, 20*(2), 295–307.

UIA. (2014). Number of international organizations by type. Edition 52, 2015/2016 (data collected in 2014). *Union of International Associations*. Available at https://uia.org/sites/uia.org/files/misc_pdfs/stats/Number_of_international_organizations_by_type_2014.pdf.

UNFCCC. (2018). Decision -/CMA.1, matters relating to Article 14 of the Paris Agreement and paragraphs 99–101 of decision 1/CP.21 – Advance unedited version. *United Nations Framework Convention on Climate Change*. Available at https://unfccc.int/documents/187714.

Verbruggen, P. (2014). Gorillas in the closet? Public and private actors in the enforcement of transnational private regulation. *Regulation & Governance, 7*(4), 512–532.

Vrabko, M., et al. (2019). *Správne právo procesné: Všeobecná časť* (2nd ed.). C. H. Beck.

Walker, B., et al. (2004). Resilience, adaptation and transformability in social-ecological systems. *Ecology and Society, 95*(5), 1-9.
Walker, B., et al. (2006). A handful of heuristics and some propositions for understanding resilience in social-ecological systems. *Ecology and Society, 11*(1), 1-15.
Wang, S. (2003). The problems of state weakness. *Journal of Democracy, 14*(1), 36–42.
WB. (1992). *Governance and development*. World Bank.
Weber, M. (1968). Economy and society. In G. Roth (Ed.), *Wittich, C*. Bedminster.
Whitacre, J. M. (2012). Biological robustness: Paradigm, mechanisms and systems principles. *Frontiers in Genetics, 3*, 1–3.
Wright, M. O., Masten, A. S., & Narayan, A. J. (2013). Resilience processes in development: Four waves of research on positive adaptation on the context of adversity. In S. Goldstein & R. B. Brooks (Eds.), *Handbook of resilience of children* (pp. 15–37). Springer.
Zafarullah, H., & Rahman, R. (2008). The impaired state: Assessing state capacity and governance in Bangladesh. *International Journal of Public Sector Management, 21*(7), 739–752.
Zhou, M. (2012). Participation in international human rights NGOs: The effect of democracy and state capacity. *Social Science Research, 41*(5), 1254–1274.

Chapter 4
Spatial Determinants of Tipping Point Governance: Beyond Stakeholder Agency

In the preceding section, a literature review regarding social tipping points was presented along with an adoption of the definition of social tipping points that will be utilized throughout this monograph. It is argued that tipping points should be understood as a social–ecological concept. Within this conceptualization, social tipping points represent a threshold, which, once crossed, triggers an abrupt and nonlinear change in the social–ecological system that is self-reinforced by a positive feedback mechanism. In the end, these nonlinear vortexes tend to lead to irreversible changes while shifting the social–ecological system from one equilibrium to a completely new one (e.g., Milkoreit et al., 2018; Fig. 2.1). Based on this conceptualization, a conceptualization of inter-dynamics between governance, resilience, and social tipping points was developed from the stakeholders' agency standpoint. Primary determinants of the resilience vis-à-vis social tipping points are according to this conceptualization the respective capacity (administrational, legal, infrastructural, fiscal, and military) of the state and non-state actors. Although the framework is openly functionalist and state-centric, the agency of non-state actors (local, private, and transnational) can augment the state capacity and resilience when facing side effects stemming from abrupt changes in the social–ecological system. Albeit this represents an important part of the argument of this monograph, it is not the whole story as we also need to take into account geographical constraints.

Naturally, many factors have an influence on resilience, and it may be futile to aspire to an exhausting account of all the features that make some places resilient while others do not (Boschma, 2015). However, given this monograph deals with the governance of social tipping points in the European Union (EU) peripheral countries, a spatial component needs to be taken into account. In the first part of this section, we demonstrate that there is a heterogeneous distribution of capacities and vulnerabilities with regard to resilience across the EU countries, pointing our attention toward the spatial and geographical aspect of resilience vis-à-vis social tipping points and why is it important to analyze the capacities from the perspective of the EU periphery. Simply put, there is a discrepancy between how different regions,

J. Szabó et al., *Governance of Social Tipping Points*, Key Challenges in Geography,
https://doi.org/10.1007/978-3-031-47413-2_4

especially the periphery and core of the EU, vary in terms of their resilience toward exogenous nonlinear shocks (e.g., Sensier et al., 2016). Given the state capacities differ substantially across the EU countries, it is reasonable to assume their resiliencies toward the social tipping points will vary as well, which is also supported by some studies (e.g., Prokkola, 2019). Subsequently, we dive into the determinants of the heterogeneity of spatial distribution of resilience within the EU that can help us explain this puzzle. Drawing from the concepts developed in the field of spatial economics, namely with regard to the literature on regional economic resilience, it is argued in this section that there are three main determinants of regional resilience going beyond the stakeholders' agency – compositional, collective, and contextual factors. As a result, this section aims to conceptualize how tipping point governance and resilience are against tipping points conditioned on spatial determinants within the EU geographical space. This argument alone is not groundbreaking as it has been part of the spatial economics literature for a long time (e.g., Crespo et al., 2013). However, in combination with the stakeholder agency conceptualization of the social tipping point governance, the spatial component will allow us to capture the determinants of the resilience of the EU's periphery against the abrupt changes within the social–ecological systems more precisely. Furthermore, this bivariate (state-centric) stakeholder agency and spatial component combined in one framework result in an otherwise counter-logical argument that not only do the peripheral states need to augment the EU's capacity, which had been conceptualized as a transnational non-state actor sui generis, but the EU and the geographical core member states of the EU need to fortify the EU periphery in order to enhance their respective resiliencies (see further Sect. 7.5).

4.1 EU Core–Periphery Dichotomy: A Spatial Distribution of Vulnerabilities and Capacities

Before we delve into the regional determinants of resilience, let us briefly examine the distribution of vulnerabilities and resilience capacities across the EU countries to assess whether the core–periphery dichotomy with regard to resilience persists within the EU. Following the COVID-19 crisis, the concept of resilience became a new compass for EU policies and its member states with the aim of conducting strategic foresight within the EU (EC, 2020c). In the foresight, resilience is proxied via four dimensions: digitalization, geopolitics, green transition, and social-economic development, whereas the holistic assessment of resilience is estimated using a set of indicators within the so-called resilience dashboards (EC, 2020b). Every indicator aims to assess the relative vulnerabilities or capacities of each EU member state and identify areas that are in need of some policy action. Vulnerabilities refer to elements of a country's system that can be disproportionally hit in the case of shock or structural change. Capacity, on the contrary, represents a country's features that are crucial for coping with potential shocks or structural changes. The synthetic indicators reflect the four aforementioned dimensions which, in turn, cover the megatrends facing the EU today. The list of all indicators for each respective area can be found in the annex to the resilience dashboard report (EC, 2021a, b).

Let us start with the digital dimensions. Digital resilience revolves around four areas, among which three are thematic – industries, personal, and public space – and one is cross-sectional and cut across all the other three – cybersecurity. These areas are in line with the Commission's plans for Europe's digital decade (EC, 2021a, b). The distribution of digital vulnerabilities across the EU countries is displayed in Fig. 4.1. The figure clearly shows that the most vulnerable countries in terms of digital challenges are countries from the eastern and southern parts of the EU periphery, mainly Poland, Hungary, Croatia, Romania, and Bulgaria. High vulnerabilities are mostly within the industry, personal, and public space areas. Not cybersecurity though, as high values for vulnerability in cybersecurity areas are present, paradoxically, in countries such as Belgium, Luxemburg, and Sweden. Factors contributing to the digital vulnerabilities of various EU member states are escalating US–China technological rivalry, the divide in Internet coverage between urban and rural areas, scarcity of data, and, most importantly, the onset of AI and its potential impact on the job markets (EC, 2020c).

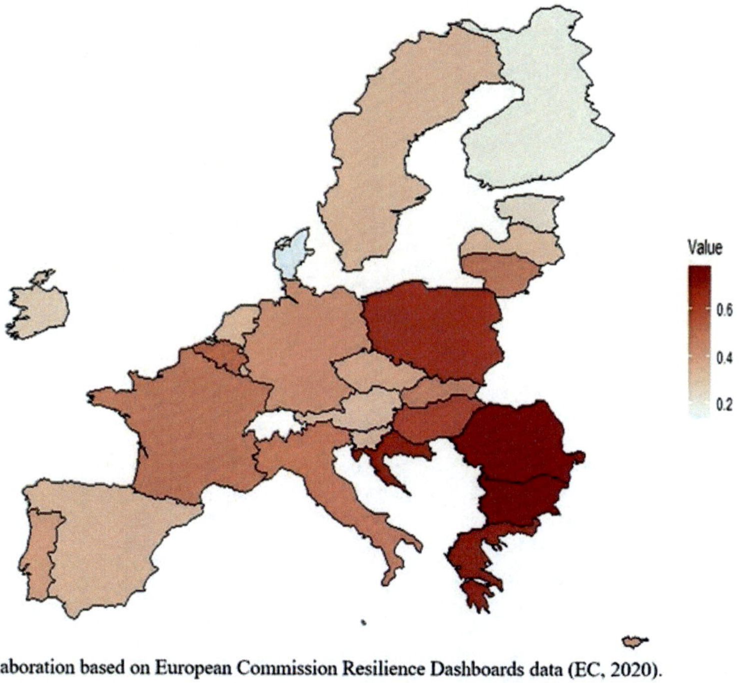

Fig. 4.1 Digital vulnerabilities index of the European member states, 2020. (*Source*: Own elaboration based on European Commission Resilience Dashboards data (EC, 2020a))

Next, the distribution of the geopolitical vulnerabilities is displayed in Fig. 4.2. The notion of geopolitical resilience in this context refers to the attempts to bolster Europe's "open strategic autonomy" as the EU's commitments to open and fair trade and multilateralism as well as the reduction of the EU's dependency on technological and value chains (EC, 2020a). The synthetic geopolitical vulnerabilities index stems from four areas – (1) financial globalization and concentration on foreign direct investment, (2) dependencies and securities with regard to raw materials and energy supplies, (3) security, defense, and demography, and lastly (4) value chains, trade partners, and economic openness (EC, 2020c). The main challenges underpinning the geopolitical vulnerabilities within the EU are the erosion of multilateralism and the global financial system retreating under the pressure of national interests, shift of power away from the Western world, global insecurity and lack of unity in foreign and security policies, overall reliance on non-EU suppliers of essential raw materials, as well as migration flows (EC, 2020c). The highest levels of vulnerabilities recorded were in Ireland and Cyprus, mainly due to excessive reliance on foreign capital and trade openness. This observation is not surprising given the economic model of these countries is built on attracting foreign capital to deregulated and relatively anonymous financial systems and predatory low corporate taxes (Bohle & Regan, 2022). A critical vulnerability stemming from the high trade

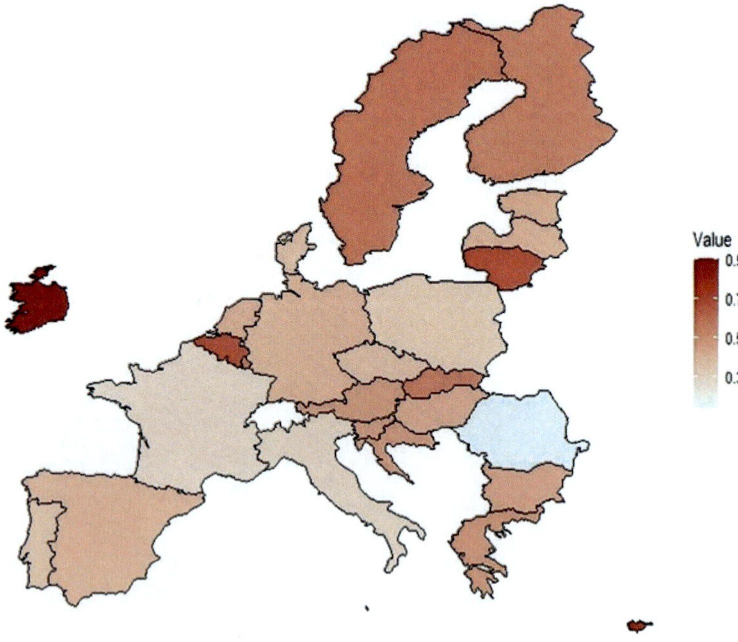

Source: Own elaboration based on European Commission Resilience Dashboards data (EC, 2020).

Fig. 4.2 Geopolitical vulnerabilities index of the European member states, 2020. (*Source*: Own elaboration based on European Commission Resilience Dashboards data (EC, 2020))

openness is, however, recorded throughout the EU's eastern periphery, namely in Estonia, Lithuania, Slovakia, and the Czech Republic. Nordic countries such as Sweden and Finland, but also Lithuania, evince high levels of vulnerabilities in raw material and energy supplies dependency, contributing to their relatively high overall levels of geopolitical vulnerability.

Green resilience generally refers to the EU's ambitions to reach climate neutrality by 2050 while mitigating the effects of climate change and adapting to some of its irreversible impacts as well as restoring the capacity of ecological systems in order to secure sustainable life within planetary boundaries (EC, 2020c). That is why the green resilience dashboard is structured along three areas: (1) climate change mitigation and adaptation, (2) sustainable use of resources, and (3) ecosystems, biodiversity, and sustainable agriculture. These goals are broadly aligned with the policy areas of the European Green Deal (EC, 2019). Current megatrends negatively influencing the green areas within the EU are too numerous to be counted (see also IPCC, 2022a, 2022b). Among the major climate vulnerabilities within the EU are currently according to the European Commission extreme weather events, hazardous substances, outsourcing of pollution to non-EU countries with lower environmental standards, immense exploitation of renewable and nonrenewable natural resources, and even over-urbanization and poverty (EC, 2020c). The distribution of green vulnerabilities across the EU countries is displayed in Fig. 4.3. The EU

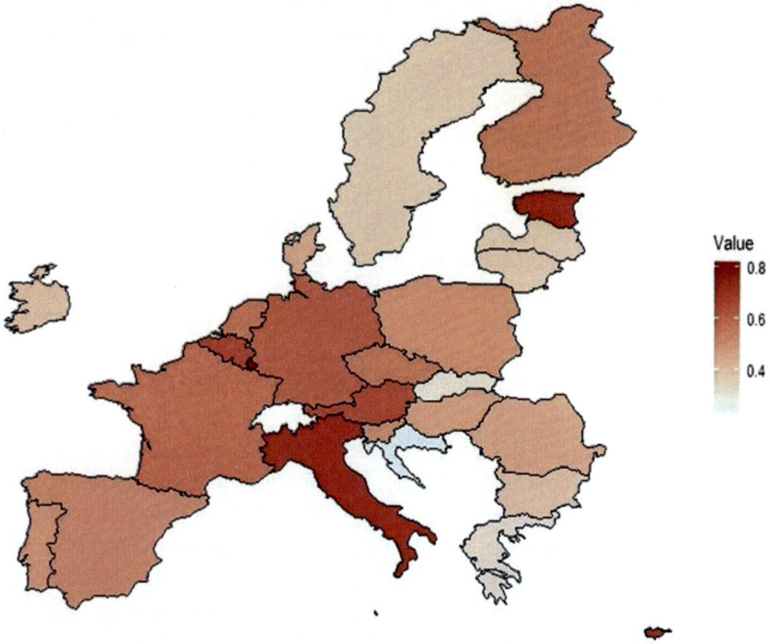

Source: Own elaboration based on European Commission Resilience Dashboards data (EC, 2020).

Fig. 4.3 Green vulnerabilities index of the European member states, 2020. (*Source*: Own elaboration based on European Commission Resilience Dashboards data (EC, 2020))

country with the highest levels of green vulnerability is, paradoxically, Luxembourg, followed by Estonia, Italy, and Cyprus. In spite of being unequivocally core EU member state, one of the founding six countries in fact, Luxembourg remains the country with the highest greenhouse gas emissions per person in the EU, is highly dependent on energy imports for final consumption, and has one of the lowest shares of renewable energy in the EU (EC, 2022).

Lastly, let us turn our attention to the social and economic dimension of resilience. In essence, this dimension covers "the ability to tackle economic shocks and achieve long-tern structural change in a fair and inclusive way" (EC, 2020c). In addition to economic growth, the European Commission also puts an emphasis on social and regional cohesions, support for the most vulnerable cohorts in society, and demographic sustainability, fully in line with the European Pillar of Social Rights. Thus, the social-economic vulnerabilities index covers three areas: (1) economic and financial stability and sustainability, (2) health, education, and work, and lastly (3) inequalities and social impact of the green and digital transitions.

Following the COVID-19 crisis, major EU-wide social-economic vulnerabilities stemmed from the disruptions in health and social care sectors, health effects of the confinement measures such as loneliness and mental health challenges, widening of the economic, gender, ethnic, and socioeconomic inequalities, and harsh impacts of the confinement measures on numerous economic sectors and labor markets, not to forget the additional pressure on the already shrinking Europe's population (EC, 2020c). Figure 4.4 reveals that the most vulnerable countries from the standpoint of the social-economic vulnerabilities index are southern, southeastern, and eastern peripheral EU countries, namely Greece, Romania, Italy, Spain, Bulgaria, and Hungary. Medium-to-low vulnerabilities are mostly associated with the EU-core countries, such as Austria, Germany, or Luxembourg. The prevalence of the southern peripheral countries among the most socially and economically vulnerable EU countries is not surprising given these countries underwent severe economic upheavals during the European sovereign debt crisis (e.g., Matthijs & Blyth, 2015). Unsurprisingly, the highest economic, financial, and sustainability vulnerabilities were recorded in the case of Greece, Cyprus, and Portugal, followed by Italy and Spain. With regard to health, education, and work, very high vulnerabilities are also recorded in Romania, Slovakia, and Croatia. The same applies to inequalities and social impact of the transitions, which are again dominated by Romania, Bulgaria, Slovakia, and Hungary (EC, 2020b).

Based on the vulnerability indices in four resilience areas – digital, geopolitical, green, and social-economic – we can see that the most vulnerable areas are countries on the geographical periphery of the EU, namely from eastern (Visegrad countries), southeastern (Bulgaria and Romania), and southern (Cyprus, Greece, Spain, Portugal, and Italy) peripheries. This observation is correlated with the previously stated hypothesis that state strength is a function of state capacity (Fig. 3.2). Let us, therefore, turn to the second side of the same coin – the resilience capacity, or the country's capacity for coping with potential shocks or structural changes. (EC, 2020b). Table 4.1 displays a review of the respective resilience capacities (digital, geopolitical, green, and social-economic) of the EU member states and the EU-27

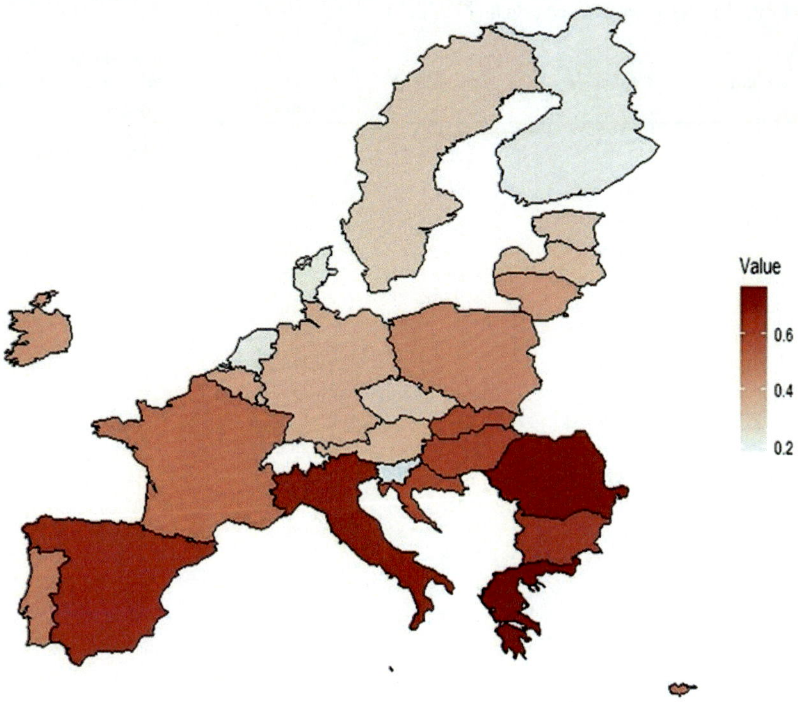

Source: Own elaboration based on European Commission Resilience Dashboards data (EC, 2020).

Fig. 4.4 Social-economic vulnerabilities index of the European member states, 2020. (*Source*: Own elaboration based on European Commission Resilience Dashboards data (EC, 2020))

average (shaded row), as well as the unweighted average of the four synthetic resilience capacities indices. The values of the average capacities index are in descending order with the aim of better visualizing the EU countries with the lowest resilience capacities. It is more than obvious that the lower positions of the ranking are occupied by the aforementioned peripheral countries. It is especially troubling when we realize that some of the countries, such as Romania, achieve close to 20% of the overall resilience capacity when compared with the first countries in the ranking – the Netherlands, Belgium, and Luxembourg. Also, Romania attains only around 30% of the EU27 average capacities. Similarly, Bulgaria and Greece are not that far from Romania in this regard.

It comes as no surprise that the distribution of the respective resilience capacities seems to be somehow dependent on the economic development of the EU member states as the EU member states with the lowest capacities seem to overlap the member states with the lowest levels of economic development (see further Sect. 7.5 and Table 7.1). Again, this observation does not seem to be very surprising given the well-documented relationship between resilience and economic development (e.g.,

Table 4.1 Resilience capacity indices of the European Union member states, 2020

Country	Digital Capacities Index	Geopolitical Capacities Index	Green Capacities Index	Social-Economic Capacities Index	Average Capacities Index
Netherlands	0.88	0.75	0.76	0.82	0.80
Belgium	0.90	0.79	0.67	0.80	0.79
Luxembourg	0.75	0.86	0.82	0.73	0.79
Sweden	0.85	0.57	0.74	0.90	0.77
Finland	0.93	0.57	0.63	0.84	0.74
Denmark	0.90	0.35	0.82	0.86	0.73
Austria	0.68	0.61	0.86	0.78	0.73
Estonia	0.92	0.65	0.67	0.56	0.70
Ireland	0.72	0.78	0.52	0.76	0.70
Germany	0.73	0.48	0.81	0.75	0.70
EU27	**0.66**	**0.48**	**0.75**	**0.62**	**0.63**
France	0.64	0.51	0.60	0.74	0.62
Slovenia	0.56	0.62	0.66	0.65	0.62
Hungary	0.74	0.53	0.54	0.53	0.58
Malta	0.85	0.67	0.33	0.45	0.57
Portugal	0.68	0.45	0.63	0.51	0.57
Croatia	0.67	0.51	0.61	0.45	0.56
Czechia	0.48	0.51	0.63	0.56	0.54
Italy	0.36	0.52	0.56	0.67	0.53
Lithuania	0.63	0.54	0.42	0.48	0.52
Poland	0.45	0.55	0.50	0.55	0.51
Slovakia	0.57	0.53	0.48	0.40	0.49
Cyprus	0.48	0.52	0.36	0.56	0.48
Latvia	0.52	0.42	0.48	0.28	0.43
Greece	0.36	0.39	0.42	0.30	0.37
Bulgaria	0.27	0.50	0.31	0.21	0.32
Romania	0.27	0.27	0.28	0.15	0.24

Source: Own elaboration based on European Commission Resilience Dashboards data (EC, 2020b)
Note: The average capacities index represents an unweighted average of the digital capacities, geopolitical capacities, green capacities, and social-economic capacities indices. The values are displayed in descending order based on the average capacities index

Dincecco & Katz, 2014). Furthermore, when we examine the associations between the gross national income (GNI) and the average of the four synthesized resilience capacity indicators, we immediately notice a strong and positive linear relationship between the two across the EU member states (Fig. 4.5). In other words, the higher the national income of an EU member state is, the higher the resilience capacity it has. This is a useful observation as it might partially explain the spatial component of tipping point governance.

To summarize, the data regarding the degree of vulnerability and capacity in digital, geopolitical, green, and social-economic areas suggest the (relative) lack of resilience can be found in part of the EU eastern (Visegrad countries, some Baltic states), southeastern (Bulgaria and Romania), and southern (namely Greece, Spain,

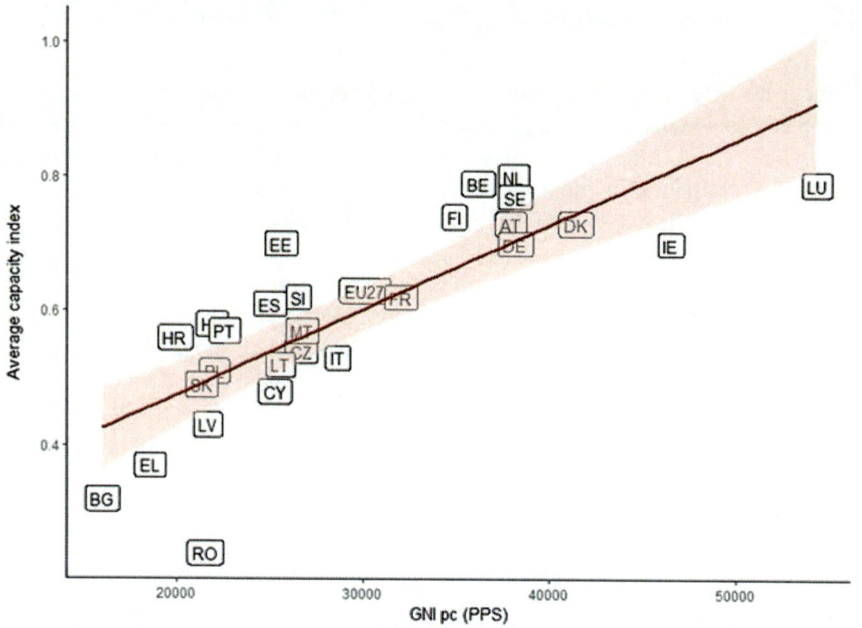

Source: Own elaboration based on European Commission Resilience Dashboards data (EC, 2020b) and Eurostat data (NAMA_10_PP). *Note*: GNI is in per capita form and in purchasing parity.

Fig. 4.5 Association between national income and resilience capacity, EU countries 2020. *Note*: GNI is in per capita form and in purchasing parity. (*Source*: Own elaboration based on European Commission Resilience Dashboards data (EC, 2020b) and Eurostat data (NAMA_10_PP))

Portugal, and Italy) peripheries. Caraveli (2017) confirms this geographical distribution of core–periphery patterns across the EU. Historically speaking, the West–East divide was the most prominent, especially following the big bang EU enlargement in 2004 which accentuated the peripheral position of countries such as Czechia, Hungary, Poland, Slovakia, the Baltic states, as well as Romania and Bulgaria. Amid the euro debt crisis, however, the North–South divide started to dominate the core–periphery relations (Matthijs & McNamara, 2015), and the subsequent economic deterioration of these countries finally cemented the current core–periphery relationships within the EU. What is more important from our perspective is that this unequal distribution of economic capacities mirrors itself in the countries' capacity to be resilient. It was clearly demonstrated that there is a correlation between the resilience capacity and national income, and there is a positive trend between the two. The higher the income an EU country has, the more resilient it becomes. Additionally, as Healy and Bristow (2019) rightly point out, even the presence of the external border in the EU (eastern and southern) periphery influences the overall resilience which is a pivotal trend that, however, tends to be downplayed.

4.2 Regional Economic Resilience: A Brief Glance

In the real world, no exogenous shock – a shock originating outside the system – is nationally, regionally, or globally neutral, and its impacts are almost always spatially conditioned (Martin & Sunley, 2015). In addition to systemic disturbances, regionally originating disruptions are also very frequent. In any case, the spatial component of resilience toward disturbances is very important. To help us to better understand its nuances across the EU countries, we draw from the literature on *regional economic resilience* (RER). In general, as is often the case in various academic fields, the RER has no unified definition. Martin and Sunley (2015: 3) note that "different authors employ different definitions and descriptions, sometimes even invoking the term without giving any precise interpretation." Most of the time, they say, the RER has no generally accepted methodology for the operationalization of the concept and not even a proper theory behind it. Hill et al.'s (2008: 4) definition is, for instance, as follows: "the ability of a region … to recover successfully from shocks to its economy that either throw it off its growth path or have the potential to throw it off its growth path." Martin and Sunley (2015: 13) define regional economic resilience as "the capacity of a regional or local economy to withstand or recover from market, competitive and environmental shocks to its developmental growth path" and "if necessary, by undergoing adaptive changes to its economic structures and its social and institutional arrangements, so as to maintain or restore its previous developmental path, or transit to a new sustainable path." Based on their conceptualization, regional resilience as a process is determined by five basic elements: (1) vulnerability (sensitivity), (2) nature of the shock, (3) initial resistance to the shock, (4) robustness of the stakeholders' response, and (5) recoverability. The authors further emphasize the recursiveness of the regional economic resistance as a shock (e.g., social tipping point), and the response to the shock and its side effects may further influence the region's resistance and robustness to subsequent shocks (see also Simmie & Martin, 2010). Even though they mostly focus on "economic shocks," Martin and Sunley (2015: 14) acknowledge that much of their RER can also apply in the case of disturbances caused by natural and environmental disruptions. What is important according to their opinion is that different shocks originate at different spatial scales (global-national-local) and most of them are abrupt, unexpected, and sudden events, which resembles our conceptualization of social tipping points (Ibidem).

The literature on RER has its origins in regional business cycles and fluctuations studies (e.g., Vining, 1946; Neff, 1949). Most of the attention in the RER literature is dedicated toward finding out why certain regions are more sensitive to the business cycles than others, the degree of synchronicity across regions, as well as what are the determinants of regional variation in terms of the sensitivity (Martin et al., 2016). The interest in researching RER, however, recently sprang out of spatial economics in the aftermath of the Great Financial Crisis (GFC) of 2008 and 2009 when many authors inquired into why certain areas reacted differently to the common exogenous shock (e.g., Sensier et al., 2016; Di Caro, 2017; Faggian et al.,

2018; Wang & Wei, 2019). These studies have shown that the regional economies experienced differences in their ability to recover from the GFC (Hadjimichalis & Hudson, 2014). This further accelerated the interests of the economist regionalist in the notion of the RER (Bristow & Healy, 2018). Groot et al. (2011), studying the sectoral composition of European countries on the NUTS1 (state) and NUTS2 (regional) levels, conclude that variation in the sectoral composition contributes to the variety of resilience toward the global recession both at the country and regional level across Europe. It is argued that countries and regions with higher shares of cyclical sectors (e.g., manufacturing) were more affected by the global recession than the other countries and regions. Similarly, the negative influence of manufacturing on regional economic resilience was found in Greece and Spain, respectively (Cuadrado-Roura & Maroto, 2016; Giannakis & Bruggeman, 2017). Moving beyond "structural" aspects of the RER, Obschonka et al. (2016) inquire into whether also "macropsychological" factors have an impact on regional economic resilience. Using a large psychological dataset from the United States and Great Britain, the authors examine the relationship between regions' levels of psychological traits and their economic growth during the GFC. It is concluded that, in both countries, "more emotionally stable regions and regions with a more prevalent entrepreneurial personality makeup showed a significantly lower economic slowdown" (Ibidem).

Another strand of literature regarding the RER started to emerge in the context of the COVID-19 crisis. Examining the Northeast industrial regions in China and their response to the COVID-19 pandemic, Hu et al. (2022) demonstrate a significant difference in how these economic regions responded to the GFC crisis and COVID-19. It is argued that in contrast to the GFC, economic openness, structure of industry, overall variety, and city size negatively influence regional resilience. This might be due to the fact that contrary to the GFC, COVID-19 was first and foremost a political and social crisis with strong interference by the authoritarian government. Similarly, Cheng et al. (2022) analyze the economic resilience of Chinese cities using GIS. The authors also conclude that compared with the GFC, the COVID-19 pandemic had a more widespread impact. Other authors also put an emphasis on the role of governance and leadership in enhancing regional economic resilience vis-à-vis the COVID-19 pandemic in China, Kunshan. Via state-led adaptive resilience strategies (stabilizing labor supply, mitigating supply-chain disruptions, alleviating financial strains, and reconfiguring market orientations), the local state was enabled to align diverse stakeholders and thus enhance regional economic resilience (Wu et al., 2023).

In recent years, the literature on the RER produced a substantial number of articles formulating a variety of resilience indicators and frameworks to better understand whether a particular region is sufficiently resilient or not, which reflects rising demand from the side of policymakers keen to assess the resilience of their political economies (e.g., CLES, 2010). Sensier et al. (2016) operationalize the concept of resilience and develop a method for measuring "revealed" resilience using GDP and employment time series across Europe during recession allowing them to develop an analysis of the shocks in spatial and temporal dynamics. Especially, the authors' approach allows for "dating the onset of the effects of a shock at regional level."

Drawing upon data from the EC's Regional Innovation Scoreboard, Bristow and Healy (2018) construct a novel measure of the European regions' resilience to the GFC. In particular, the authors analyze the relationship between regional innovation capacity and regional economic resilience. It is demonstrated that "the capacity for innovation within a [European] region is strongly related to its propensity to be resilient to economic shocks" (Ibidem). Similar results were also demonstrated by Filippetti et al. (2020). From the other side of the globe, Wang and Wei (2019) propose a new measure of resilience based on the theories of simple harmonic motion and regime switching in order to examine the determinants of resilience among China's regional economies to the GFC. The authors conclude that the industrial diversity, human capital stock, trade openness, and financial liberalization seem to improve regional economic resilience. A new composite policy tool to measure territorial resilience was also developed by Pontarollo and Serpieri (2020a). The authors identify the main dimensions of RER and synthesize them into a composite indicator in order to categorize the resilience degree across the EU regions into clusters. It is demonstrated that national resilience trends dominate in the "old" member states, whereas more heterogeneous spatial patterns are recorded in the "new" member states that joined the Union after 2004.

Be it various works analyzing the spatial distribution of the effects stemming from the GFC, response to COVID-19 in numerous Chinese regions, or multiple indicators and frameworks to measure and assess regional economic resilience, it is clear that the literature on RER provides some answers regarding the disproportionate lack of resilience at the EU periphery (eastern and southern) and its relative "abundance" in the core EU countries. Therefore, in spite of the RER having its focus on the region as its basic unit of analysis, the majority of its principles can be applied to the country-level differences as well, especially within the EU context having quasi-constitutional features (see Sect. 3.5.4) and institutionalized single market posing (almost) no barriers to the free flow of goods, capital, and information (Egan & Guimarães, 2017). As a result, even though this monograph is mostly focused on the multilevel governance of social tipping points in social–ecological systems from the state-centric perspective, the spatial determinants of resilience in the EU can be borrowed from the RER. Furthermore, the concept of resilience within the field of spatial economics is frequently used with different meanings and through various units of analysis (e.g., Modica & Reggiani, 2015), and within the EU context, the geography of resilience is clearly influenced by national patterns (e.g., Bristow et al., 2014).

4.3 Determinants of Regional Economic Resilience

In the previous section, it was demonstrated the RER has a place in explaining the dispersion of resilience across the EU countries. According to Martin and Sunley (2015), a central question about resilience is why it depends on geography and what are its determinants. Most importantly, the determinants that enhance a region's

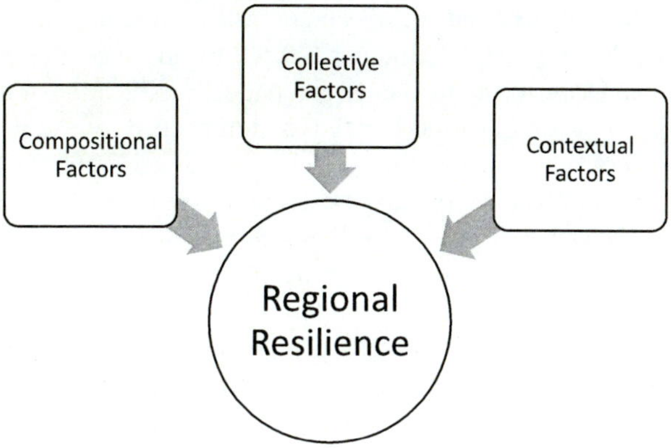

Source: Own elaboration based on Grabner (2016: 28).

Fig. 4.6 Determinants of regional resilience. (*Source:* Own elaboration based on Grabner (2021: 28))

resilience during disturbances tend to average be the same that enhance their growth potential and competitiveness during stable times (e.g., Fratesi & Rodríguez-Pose, 2016; Di Caro & Fratesi, 2018). The determinants of RER are both complex and dynamic and are highly dependent on the intersections between factors within regions as well as external forces and relations (Bristow & Healy, 2014). Furthermore, they are of a necessary "multi-scalar nature." Sutton and Arku (2022) conceptualize the determinants of RER are products of internal and external key actors, which are in interaction with the region's economic structure and its economic, social, and political environment. Borrowing concepts from behavioral psychology and health studies, Martin and Sunley (2015) argue that the RER is produced by a combination of three sets of factors: compositional, collective, and contextual (Fig. 4.6). Compositional factors refer to structural aspects, collective factors to socioeconomic networks, whereas contextual factors cover the institutions. These factors are highly interdependent, need to complement each other, and mirror the underlying regional perceptions. Furthermore, they shape the reactors of heterogeneous agents and might explain the difference in response to resilience against social tipping points between the EU core and periphery (Ibidem). In the subsequent section, we first theoretically conceptualize the three determinants of regional resilience according to Martin and Sunley (2015) and then proceed to assess how these might allow us to comprehend the diversity of resilience across the EU countries vis-à-vis social tipping points.

4.3.1 Compositional Factors

In general, compositional factors are closely related to the economic structure – sectoral composition – of a regional unit, and it is widely accepted that they are pivotal to regional resilience. These mainly refer to the makeup of the regional or local economy. In a nutshell, different sectors and industrial structures are affected differently by shocks. It is argued, by Grabner (2021), for instance, that regional structural diversity enhances overall resilience given the diversification of the risk among different regions in case of a shock. It is thus recommended for a regional economic unit to avoid overall dependence on a single specialized sector to become more resilient. To put it simply, the more diverse a region economically is, the more resilient it becomes (Davies & Tonts, 2010; Wang & Wei, 2019). Naturally, it does not always have to be that simple as the relationship between regional economic resilience and economic diversity may be conditioned on the magnitude of the shock. For instance, there might be an increased likelihood of exposure if the regional economic units have more diverse regional economic activity. Therefore, Boschma (2015) argues that paradoxically, a specialized region may be less vulnerable to many asynchronous shocks, but, unproportionally, more exposed to a major crisis. All in all, the whole "debate between specialization and diversity" is far from settled (Martin & Sunley, 2015: 30).

A very important aspect of the compositional factor is also the so-called *modularity*, or "a system structure in which different component subsystems or elements are only partially or weakly connected or linked, so that if one such subsystem or element is affected by a shock, the effect remains relatively contained and its diffusion throughout the whole system is minimized" (Martin & Sunley, 2015: 7). Loosely connected sectors may potentially retain exogenous shocks locally and prevent contagion to other regions, and if only one module lacks resilience, the whole system may nevertheless remain robust in the face of perturbations. The concept of modularity is closely connected to the concept of *redundancy* or a situation, in which similar modules can replace each other when one does not endure the pressure. This modus operandi, however, happens quite rare, and a more common mechanism is a system consisting of diverse components with complementary roles whereby a specific systemic function can be carried out by alternative means available (Ibidem). Therefore, not only the structural variety is important but also the extent to which that variety confers modularity and redundancy is important. The modality puts a strain on the spread of shock across the (economic) structures, whereas the redundancy may increase the possibility of reorienting that shock toward the more resilient and robust sectors. Both are thus vitally important. Naturally, this may be a double-edged sword as regional connectedness and complementarities in industrial structures may also have the potential to enhance the adaptability of the local economy (Boschma, 2015). For instance, Evans and Karecha (2014) suggested that regional economic resilience in Germany is partly attributable to the diverse industrial structure whereas Giannakis and Bruggeman (2015) observe that agriculture strengthened the resilience of tourism in rural Greek island regions reliant on tourism. Similar results with regard to regional economic resilience were also reported with regard to the autonomous region of Andalusia (Sánchez-Zamora et al., 2014) and European regions (Xiao et al., 2018). In our case, however, we are mostly interested in the compositional factors influencing resilience and a well-balanced approach between modularity and close connectedness of the sectors is probably the most viable alternative. In addition to modularity and redundancy as aspects of structural diversity, Martin and Sunley (2015) also borrow a concept of the "rivet effect" from ecological studies. This concept describes a systemic component or species that is of utmost importance for the functioning of the ecosystem. Accordingly, their collapse might spill over into the failure of the entire ecosystem. Then, once hit by a shock, the rivet structure (e.g., pivotal refugee camp) pulls the whole social–ecological system down with it.

Needless to say, material sector diversity alone, albeit pivotal for regional resilience, is simply not enough. Grabner (2021), for instance, also emphasizes other composition factors, namely human capital, knowledge, and demography. According to the author, human capital is central to regional development and ipso facto resilience, giving rise to innovative and high-wage industries. Within the EU, regions with the highest capacity for innovation tend to have the highest capacity for resilience (Bristow & Healy, 2018; Filippetti et al., 2020). It is well documented that well-educated population is more capable of buffering sudden economic shocks and adapting to long-term structural changes given the transferability of their skills (e.g., Rodríguez-Posé, 2013; Giannakis & Bruggeman, 2017; Di Caro & Fratesi, 2018) which mirrors itself in the resilience and adaptability of knowledge economic sectors. These sectors tend to absorb shocks at a faster rate than other sectors due to a greater propensity to create new products and improve production processes (Storper & Scott, 2009; Bristow & Healy, 2018; Di Caro & Fratesi, 2018). It is thus crucial for the regions wishing to become more economically resilient to increase the quality of human capital, knowledge economy, and innovation capacity (Bristow & Healy, 2018). Furthermore, given that people endowed with higher-than-average human capital tend to be very mobile, it is also necessary to be able to not only generate high human capital but also attract and retain highly educated people in the region with the aim of augmenting regional resilience (Martin & Sunley, 2015). The last structural aspect being crucial for the overall economic resilience is the so-called *entrepreneurism*, which states that entrepreneurs tend to enhance the economic resilience due to their adaptive capacity and capability to be innovative and develop new growth paths (e.g., Obschonka et al., 2016).

4.3.2 Collective Factors

Beyond structural structures, collective factors also need to be taken into account. The collective factors mainly refer to regional networks and their dependency on the economic relationship beyond the system boundaries determining the respective sensitivity of regions to shocks (Boschma, 2015). According to Grabner (2021), the network as a collective factor consists of organizations, crucial stakeholders, and respective nodes and relatedness between them, whereas the proximity between respective units are crucial aspects for determining the overall resilience. In particular, the author emphasizes the structural aspects of the networks. As a result, social capital and social cohesion are pivotal in improving regions' collective response to unexpected perturbations and, therefore, increase their capacities and abilities to develop new paths of recovery, which, in turn, enhances overall societal resilience (e.g., Murua & Ferrero, 2019; Sabatino, 2019). Furthermore, from a purely economic perspective, "the core/periphery structures are especially conductive to resilience" (Grabner, 2021: 32). The better the core and peripheral regions are connected, the better the transmission mechanism between the two works and enhances the overall resilience of the system as the possibility of diffusion from core to periphery and vice versa might contribute to the overall adaptability (Crespo et al., 2013). This has been empirically concluded in several works (e.g., Diodato & Weterings, 2014; Han & Goetz, 2019). In addition, networks generating and distributing knowledge play a pivotal role in adaptability (Crespo et al., 2013).

The in-system network of organizations, agents, and nodes between them is by definition an internal factor. In addition to the static network factors, highlighting the internal regional characteristics, external spatial interactions should not be disregarded, especially given all contemporary economic units are strongly interconnected via a web of economic transactions, such as flows of capital, people, goods, and services as well as information (Grabner, 2021). It is long known that the regional economies' resilience is influenced by spatial spillovers from neighboring regions. This is mostly caused by spatial dependencies (e.g., Pontarollo & Serpieri, 2020b). Within the globalized economy, collective resilience is not necessarily contained within the region, but factors tend to be exchanged among numerous regions simultaneously, frequently surpassing the political and economic boundaries of one region. Such a geographical distribution of economic transactions and a joint collaboration across regional networks can be proxied via accessibility, a measure capturing spatial interactions and connectivity (Östh et al., 2015). According to Östh et al. (2018), accessibility is a function of spatial openness. In the short run, it can make a region more vulnerable to economic shocks and potentially even enhance its propagation. In the longer run though, openness and high degree of accessibility are necessary conditions for economic development via human capital and innovation attraction, which, in turn, reinforces economic resilience.

4.3.3 Contextual Factors

Lastly, the lack of economic structures and networks alone is not sufficient to explain the missing regional resilience. To be able to holistically capture the aspects determining regional resilience, contextual factors need to be taken into account. The contextual factors revolve around the concept of institution (Martin & Sunley, 2015), formal and informal "rules of the game" (North, 1990). They are important, namely because all determinants of resilience stem from "an array of socio-economic and political-institutional factors that vary depending on economies' inherent and inherited resources, capabilities, and characteristics" (Sutton et al., 2023). Thus, specific institutional settings, quality of governance, and policies that are pursued by respective policy stakeholders substantially affect regional economic resilience. Ezcurra and Rios (2019) argue that this is mainly due to the ability of the institutional environment to drive systemic uncertainty, which subsequently also affects economic performance and resilience. Furthermore, the quality of institutions and governance structures also shape other aspects of economic resilience, such as the structural determinants and potential coordination of the responses to shocks (Sutton et al., 2023).

The importance of institutions for regional development is empirically undeniable (e.g., Rodríguez-Posé, 2013; Ezcurra & Rios, 2019). On the contrary, compared with other determinants of regional economic resilience, very little space has been given in the literature to the role of institutions. In fact, this has been a common source of criticism directed at the regional resilience literature (e.g., Hassink, 2010; Ezcurra & Rios, 2019). Adger (2000), for instance, argued that institutions are the most important determinant of regional resilience. In our opinion, it is important not to omit the role of institutions for two reasons. First, institutions are deeply path-dependent and thus crucial for their ability to mirror past legacies. According to Boschma (2015), regional resilience is strongly rooted in these past legacies, and therefore, institutions, their quality, and institutional change can provide us with information concerning the historical determinants of regional resilience. Second, institutions, be they formal (e.g., legal norms and regulations) or informal (e.g., conventions and norms of conduct), and associated social capital are key intermediaries between short-term recovery efforts and long-term development following a shock (Grabner, 2021).

4.4 Determinants of Regional Resilience vis-à-vis Tipping Points in the EU

This section demonstrates how the respective three factors of regional economic resilience – compositional, collective, and contextual – can be used to explain the distribution of resilience across the EU countries and potentially unveil what determinants beyond the stakeholders' agency may also contribute to explaining the tipping point governance within the EU geographical space. The goal of this subsection is not to provide robust empirical data confirming or refuting the determinants of resilience allocation across the EU countries but to merely point to certain trends and proven linkages driving this unequal distribution of the resilience capacities in this geographical space that are already backed by the relevant literature. Before we delve into the case studies analyzing the multilevel governance of social tipping points in the EU periphery and the role of the respective stakeholders' agency in that (Sect. 6), we consider it to be important to outline the structural and nonmaterial drivers of the lack of resilience in this part of the EU that create a bedrock for the state and non-state actors. These are, as will be shown below, *the economic diversity* and *knowledge capacity* (compositional factors), *social capital* and *trust* (collective factors), as well as *the quality of institutions* (contextual factors).

4.4.1 Compositional Factors

To summarize the composition factors determining the RER, it is first and foremost the economic structure and sectoral composition that influences the regional economic resilience. Among aspects shaping economic diversity, the modularity, redundancy, and rivet effects ought to be mentioned as well, albeit none of them alone is indispensable. Also, the demographic aspects along with the human capital and knowledge sectors are important aspects determining the RER. Let us thus start with the economic diversity aspect of economic resilience. Figures 4.7, 4.8, and 4.9 display economic concentration on the NUTS2 level across the EU countries from the perspective of employment. We opt for a regional rather than country-level approach in order to more accurately visualize the employment concentrations across the EU. This approach toward measuring economic concentration is in line with other research (OECD, 2016` Eurostat, 2019). The employment concentration is divided into three groups: industry and construction (Fig. 4.7), service sectors (Fig. 4.8), and agriculture, forestry, and fishing. Arguably, the most convincing empirical argument comes from the industry and construction perspective, where the employment concentration is the most prevalent in the industrial hub in the eastern periphery historically tied to the German automaking industry (e.g., Bohle & Greskovits, 2012). Countries such as Czechia, Hungary, Poland, and Slovakia integrated themselves into the EU's single market by "transforming themselves into one of Europe's largest transborder clusters of complex-manufacturing export industries" by attracting foreign direct investment tied to predominantly German capital (Ibidem: 138). The dependent nature institutionally and economically embedded in these countries' infrastructure of capitalism seriously diminished their resilience capacities as some of the pivotal economic decisions were taken out of the hands of the domestic stakeholders (Nölke & Vliegenthart, 2009). A similar degree of economic concentration is also visible in the southeastern EU's Bulgaria and Romania as well as in the southern peripheral regions, namely the north of Italy, Portugal, and parts of Spain.

Service sectors' concentration is visibly dominant in the more consumption-led EU growth models in the Baltics, Ireland, and other southern peripheral countries. Greece, Italy, and Spain are worth mentioning examples (Johnston & Matthijs, 2022). We should not also overlook Croatia. The Croatian case, however, is not surprising given the economy's reliance on tourism. The concentration of the Croatian economy on the tourism sector thus clearly diminishes the country's capacity to distribute the effects stemming from an unexpected tipping point hitting the country. On the contrary, we can visibly notice the core EU member states do not evince any indications of overall concentration in this sector. The same can be also said about agriculture (Fig. 4.9). With the exception of Romania and Bulgaria and parts of Poland, however, no EU regions seem to concentrate labor activity in a disproportionately large manner in agriculture. Nevertheless, some parts of Romania and Bulgaria can undoubtedly bear badly nonlinear disturbances affecting their agricultures and adapting to them.

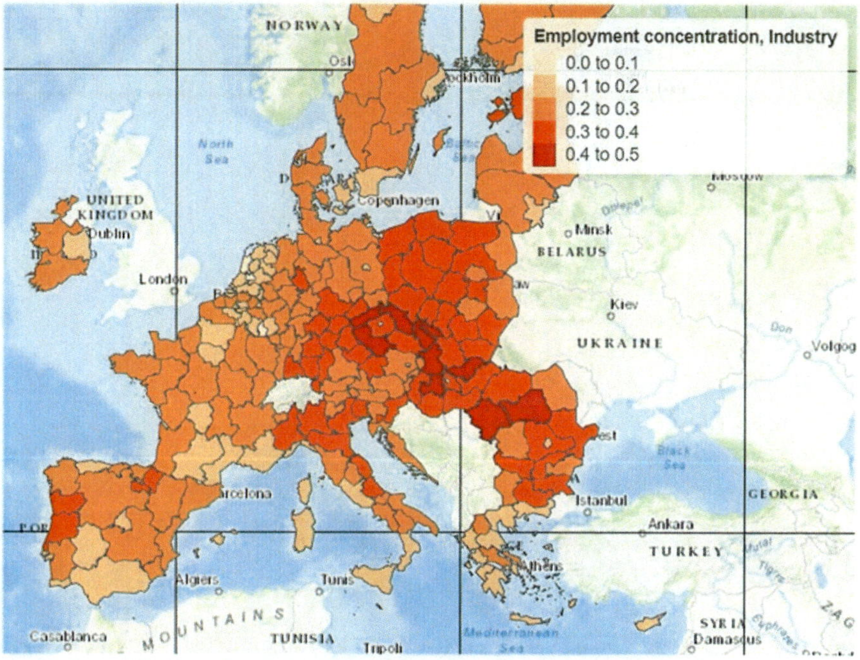

Source: Own elaboration. Map produced in R with data Eurostat data (LFST_R_LFE2EN2) from Eurostat-package http://ropengov.github.io/eurostat.
Note: NACE activities include Industry (B-E) as well as construction (F). Share of persons employed in the industry and construction on all persons employed in the economy. Map created using *tmap* package (Tennekes et al., 2022).

Fig. 4.7 Employment concentration in industry, NUTS2 regions in 2020. *Note*: NACE activities include industry (B–E) as well as construction (F). Share of persons employed in the industry and construction on all persons employed in the economy. Map created using the *tmap* package (Tennekes et al., 2022). (*Source*: Own elaboration. Map produced in R with data Eurostat data (LFST_R_LFE2EN2) from Eurostat-package http://ropengov.github.io/eurostat)

In line with Kitsos and Bishop (2018), we also take a look at the sectoral composition and diversity using the *Herfindahl-Hirschman index* (Fig. 4.10). This way, external economic relations are also taken into account. From the perspective of concentration on external demand, a few otherwise economically vulnerable economies such as Greece and Italy show unexpectedly stable and sustainable concentrations. On the contrary, the relative concentration on external demand is preconditioned on the growth model of each country (Johnston & Matthijs, 2022), and therefore, these numbers should not be automatically construed as proof of economic diversity. It is helpful to know that from the external demand perspective, the peripheral EU countries vary in terms of their market concentrations. For instance, countries such as Czechia, Poland, and Hungary – from the perspective of the eastern periphery – seem to be among the least externally diversified economies in the EU, which further contributes to their overall vulnerability and lack of resilience.

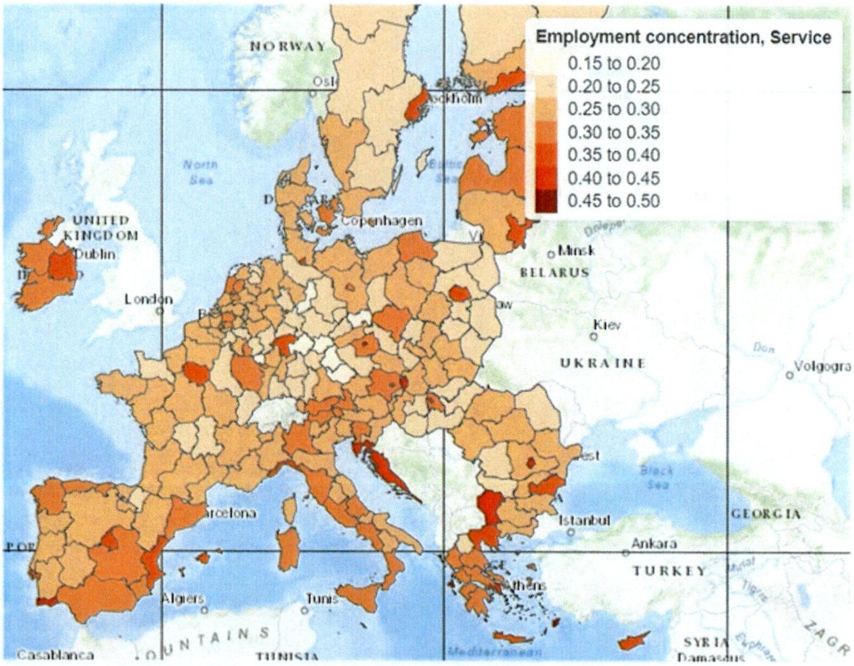

Source: Own elaboration. Map produced in R with data Eurostat data (LFST_R_LFE2EN2) from Eurostat-package http://ropengov.github.io/eurostat.
***Note*: NACE activities include Wholesale and retail trade, transport, accommodation and food service activities (G-Information and communication (J), Financial and insurance activities (K) as well as Real estate activities (L). Share of persons employed in the industry and construction on the all persons employed in the economy. Map created using *tmap* package (Tennekes et al., 2022).**

Fig. 4.8 Employment concentration in service, NUTS2 regions in 2020. *Note*: NACE activities include Wholesale and retail trade, transport, accommodation, and food service activities ((G) information and communication (J), financial and insurance activities (K), as well as real estate activities (L)). Share of persons employed in the industry and construction on all persons employed in the economy. Map created using the *tmap* package (Tennekes et al., 2022). (*Source*: Own elaboration. Map produced in R with data Eurostat data (LFST_R_LFE2EN2) from Eurostat-package http://ropengov.github.io/eurostat)

Lastly, we also need to assess the importance of the knowledge sector across the EU countries. It was argued in the previous sections that the knowledge sector is particularly important for economic resilience as sectors with higher gross value added (GVA) tend to be more resilient than traditional industries. Table 4.2 summarizes the share of knowledge sectors' shares of GVA on the overall GVA across EU member states. Once again, the relative lack of the GVA coming from the knowledge sectors in the peripheral EU countries is not surprising and has been recorded in other works as well (e.g., Rapacki et al., 2019). We can see that the Baltic States, Visegrad Four countries, as well as Romania occupy the lower half of the list, which, given the assumptions of the RER literature, automatically diminishes these countries' capacity to be resilient.

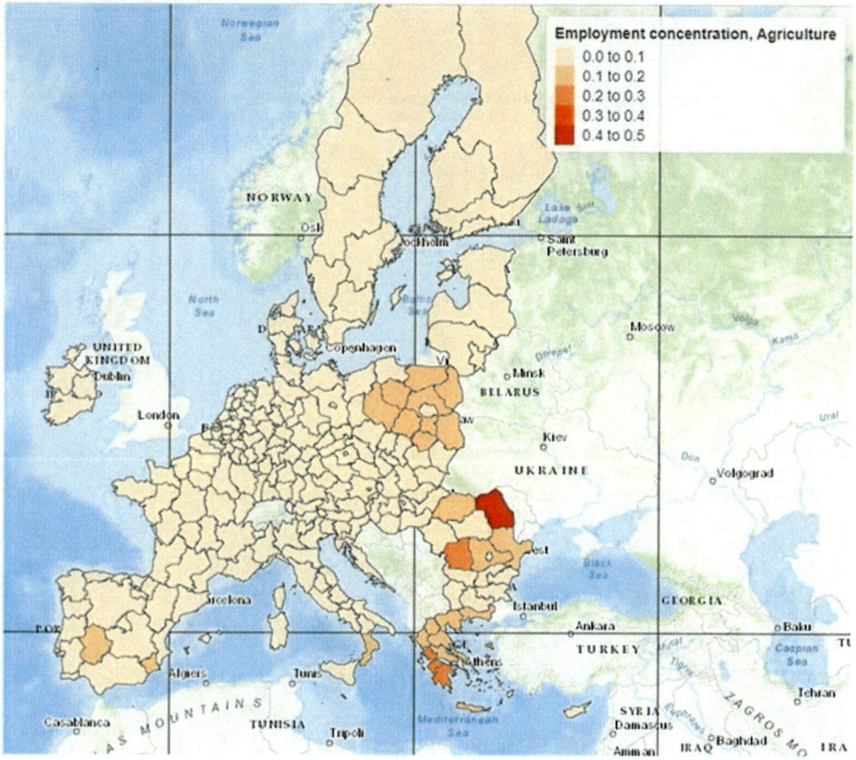

Source: Own elaboration. Map produced in R with data Eurostat data (LFST_R_LFE2EN2) from Eurostat-package http://ropengov.github.io/eurostat.
Note: NACE activities include Agriculture, forestry and fishing (A). Share of persons employed in the industry and construction on the all persons employed in the economy. Map created using *tmap* package (Tennekes et al., 2022).

Fig. 4.9 Employment concentration in agriculture, NUTS2 regions in 2020. *Note*: NACE activities include agriculture, forestry, and fishing (A). Share of persons employed in the industry and construction on all persons employed in the economy. Map created using the *tmap* package (Tennekes et al., 2022). (*Source*: Own elaboration. Map produced in R with data Eurostat data (LFST_R_LFE2EN2) from Eurostat-package http://ropengov.github.io/eurostat)

4.4.2 Collective Factors

Collective factors represent the nodes and proximity between crucial stakeholders and organizations. These networks can be, by definition, inward-oriented as well as outward-oriented. The former mostly concern the internal regional characteristics of the networks, whereas the latter also consider external spatial interactions across the regional unit's "borders." Social cohesion and social capital are the most coherent concepts that are associated with the quality of networks as determinants of economic resilience having multiple associated side effects (e.g., Rodríguez-Posé & von

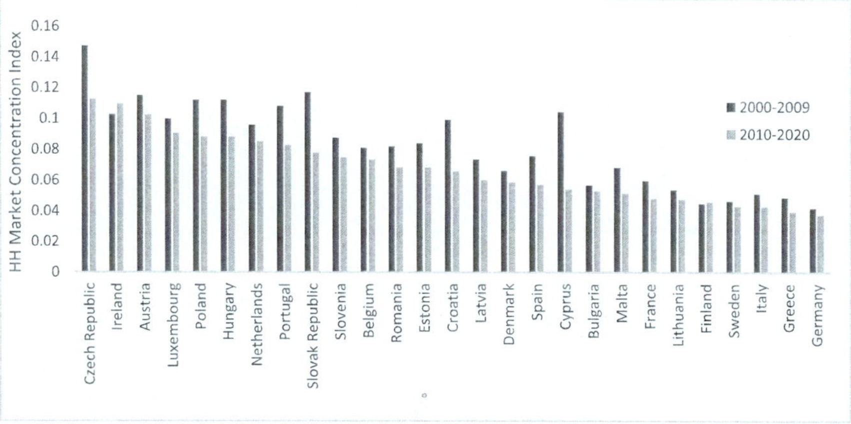

Source: Own elaboration based on World Bank data (World Integrated Trade Solution).
Note: the HH concentration index is a measure of the dispersion of trade value across an exporter's partner. A country with trade concentrated in a very few markets has an index value close to 1. On the other hand, a country with a diversified portfolio will have an index close to zero.

Fig. 4.10 Herfindahl-Hirschman (HH) concentration index across EU member states, 2000–2020. *Note*: The HH concentration index is a measure of the dispersion of trade value across an exporter's partner. A country with trade concentrated in very few markets has an index value close to 1. On the contrary, a country with a diversified portfolio will have an index close to 0. (*Source:* Own elaboration based on World Bank data (World Integrated Trade Solution))

Berlepsch, 2014). To best capture social capital, in compliance with Puškárová (2022), we rely on European Social Survey (ESS) data that contains several questions concerning trust and social cohesion. The answers from the ESS round 10 to the questions concerning trust grouped by the respondents' countries are displayed in Fig. 4.11. The respondents from the selected EU countries expressed their views regarding social cohesion, trust toward one another, and the possibility of relying on other people in times of need. It is clear that countries with greater social cohesion, on average, can be found in the EU core countries, namely Finland and the Netherlands. On the contrary, the lowest levels of trust and overall social cohesion were recorded in the eastern and southern periphery. Especially, respondents in Poland, Bulgaria, and Slovakia expressed a low tendency to trust one another, closely followed by their colleagues in southern parts of the EU, namely Portugal, Spain, and Italy.

It is thus clear that the distribution of trust and social capital also negatively tilts toward the EU peripheral member states in the east and south, which also partially explains the relative lack of economic development and indirectly influences the quality of governance in these countries (e.g., Christoforou, 2010; Graeff & Svendsen, 2013). Consequently, the transmission mechanism from low trust to relative economic lagging then metamorphosizes into a relative lack of resilience in the peripheral EU countries (e.g., Reeves et al., 2015). Bearing in mind the lack of robustness of the dataset, Fig. 4.12 nonetheless reveals an interesting correlation between trust and resilience capacities across the EU-selected member states, where higher trust levels are associated with greater resilience.

Table 4.2 Knowledge sectors' shares of gross value added across EU member states, 2015–2020

Country	*ICT*	*Finance*	*Public Sector*	*Knowledge Sector*
LU	5.73	26.55	16.07	48.35
CY	6.57	10.20	19.35	36.12
MT	7.52	8.92	17.07	33.50
NL	5.02	7.23	21.23	33.48
SW	7.77	4.25	21.27	33.28
BE	4.27	6.45	21.62	32.33
DK	4.72	5.70	21.53	31.95
FR	5.27	4.10	22.45	31.82
FI	5.88	3.15	20.58	29.62
IE	12.67	5.67	11.18	29.52
EL	3.12	5.35	20.43	28.90
EU27	**5.38**	**5.73**	**17.61**	**28.72**
HR	4.90	5.88	17.67	28.45
PT	3.70	5.10	19.47	28.27
BG	6.70	6.50	14.53	27.73
DE	4.75	4.07	18.45	27.27
ES	3.73	4.02	18.60	26.35
EE	6.30	4.45	15.57	26.32
LV	5.32	3.82	16.70	25.83
IT	3.77	5.17	16.78	25.72
HU	4.95	3.63	17.10	25.68
AT	3.68	4.20	17.63	25.52
CZ	5.83	4.17	15.13	25.13
SI	4.03	3.87	16.63	24.53
RO	6.17	3.07	14.02	23.25
PL	4.17	4.22	14.82	23.20
SK	5.07	2.82	14.83	22.72
LT	3.73	2.22	14.72	20.67

Source: Own elaboration based on Eurostat data (NAMA_10_A10)
Note: The data represent 6-year averages (2015–2020) of the knowledge sectors' gross value added on the total gross value added of the respective countries. *ICT* represents NACE's "Information and communication" (J), *Finance* "Financial and insurance activities" (K), and *Public Sectors* "Public administration, defense, education, human health, and social work activities" (O–Q). The column *Knowledge sector* represents a mere subtotal of all three columns. The EU27 is calculated as an unweighted average of all the 27 member states' values. The data are displayed in descending order based on the knowledge sector values

In spite of the limitations of the data shown above, there seems to be a well-documented relationship between trust (social cohesion) and the country's level of resilience. As a result, varying levels of social trust may influence the resilience of certain countries vis-à-vis social tipping points within the social–ecological systems. Gür (2020), for instance, analyzes the role of social trust in promoting pro-environmental behaviors with the aim of fighting climate change. The author concludes that indeed social trust increases individual willingness to take personal actions aimed at fighting climate change, even if they are time-consuming. The reason behind that is the reduction of the free-rider problem, suppression of

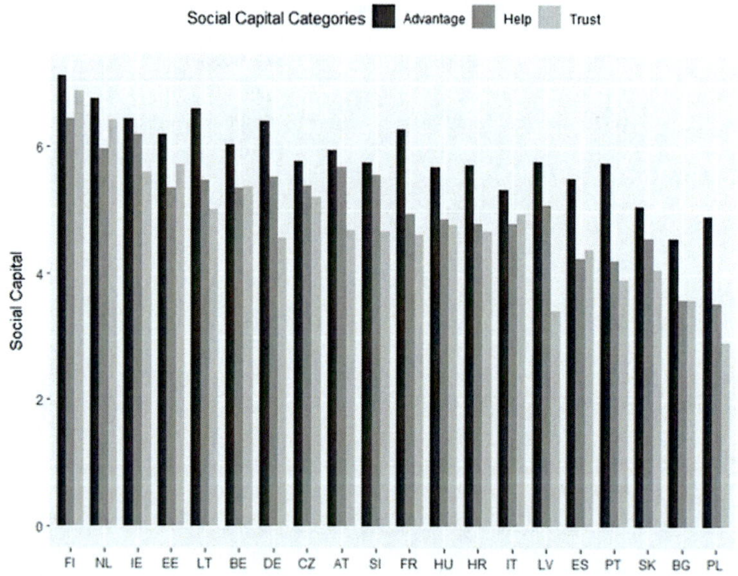

Source: Own elaboration based on European Social Survey (ESS10).
Note: Social capital categories are grouped by three questions from the ESS10 survey. *Advantage* (pplfair): "Most people try to take advantage of you or try to be fair"; *Help* (pplhlp): "Most of the time people helpful or mostly looking out for themselves" and most importantly *Trust* (ppltrst): "Most people can be trusted, or you can't be too careful". All three question range from 0 (lowest levels of trust) to 10 (higher levels of trust). "Refusal", Don't know" and "No answer" filtered out. Post-stratification weight including design weight used when computing averages per each country. Only 20 out of 27 countries' responses available in the ESS10 (Cyprus, Luxembourg, Denmark, Romania, Sweden and Malta not included).

Fig. 4.11 Social capital in selected EU member states, 2020. *Note*: Social capital categories are grouped by three questions from the ESS10 survey. *Advantage* (pplfair): "Most people try to take advantage of you or try to be fair"; *Help* (pplhlp): "Most of the time people helpful or mostly looking out for themselves"; and most importantly *Trust* (ppltrst): "Most people can be trusted, or you can't be too careful." All three questions range from 0 (lowest levels of trust) to 10 (higher levels of trust). "Refusal," Don't know," and "No answer" filtered out. Post-stratification weight includes design weight used when computing averages per country. Only 20 out of 27 countries' responses are available in the ESS10 (Cyprus, Luxembourg, Denmark, Romania, Sweden, and Malta not included). (*Source:* Own elaboration based on European Social Survey (ESS10))

opportunistic behavior, and the enhancement of cooperation. Similar arguments can also be found in the collective work edited by Cvetkovich and Löfstedt (2013). Therefore, in addition to composition factors determining the distribution of resilience and governance capacities against social tipping points within the EU, the collective factors are equally important.

4.4.3 Contextual Factors

Institutions and quality of governance are long known to be important for socioeconomic development and overall well-being (North, 1990). Based on the determinants of RER, we also know that the institutions are equally important for long-term

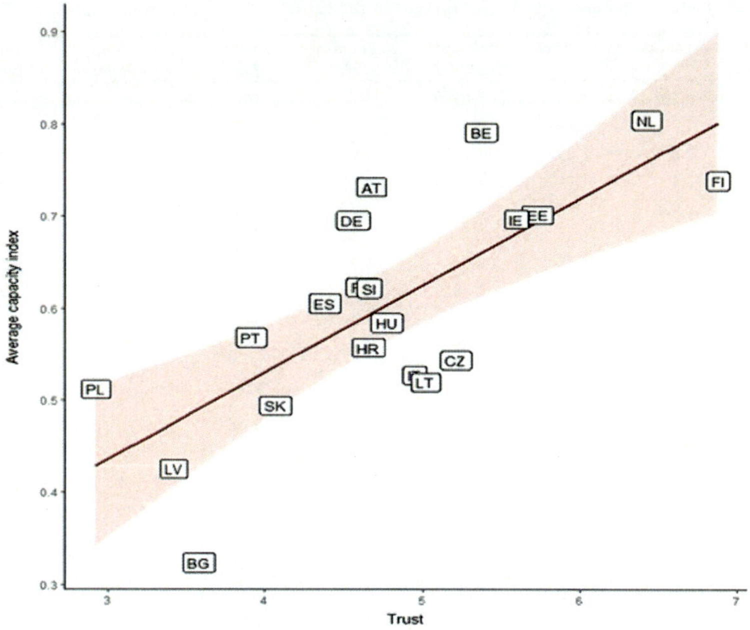

Source: Own elaboration based on European Commission Resilience Dashboards data (EC, 2020b) and European Social Survey (question ppltrst).
Note: Trust encompasses answer to the question: "Most people can be trusted, or you can't be too careful" and range from 0 (lowest levels of trust) to 10 (higher levels of trust). "Refusal", Don't know" and "No answer" filtered out. Post-stratification weight including design weight used when computing averages per each country. Only 20 out of 27 countries' responses available in the ESS10 (Cyprus, Luxembourg, Denmark, Romania, Sweden and Malta not included).

Fig. 4.12 Association between trust and resilience capacity, EU countries 2020. *Note*: Trust encompasses the answer to the question: "Most people can be trusted, or you can't be too careful" and range from 0 (lowest levels of trust) to 10 (higher levels of trust). "Refusal," Don't know," and "No answer" filtered out. Post-stratification weight includes design weight used when computing averages per country. Only 20 out of 27 countries' responses are available in the ESS10 (Cyprus, Luxembourg, Denmark, Romania, Sweden, and Malta not included). (*Source*: Own elaboration based on European Commission Resilience Dashboards data (EC, 2020b) and European Social Survey (question ppltrst))

resilience building. Unfortunately, there has been a consistent deterioration of institutional quality across regions, including weaker checks and balances and less transparency between the GFC and the COVID-19 crisis (World Economic Forum, 2020). Table 4.3 summarizes pivotal estimates of institutional and governance indicators across the EU countries, covering the quality of democratic institutions (voice and accountability and absence of violence), the quality of governance (governance effectiveness and regulatory quality), and the rule of law (rule of law and control of corruption). Unsurprisingly, when we take a look at the overall quality of institutions and governance, the bottom half of the table is once again occupied by the peripheral countries, namely from the southeast (Bulgaria and Romania), but also from the south (Greece, Croatia, and Italy) as well as east (Hungary, Poland, and Slovakia). The top places, on the contrary, are held by the core countries, especially Northern Scandinavian countries, Benelux countries, and Germany.

Table 4.3 Institutional and governance quality of the European Union member states, 2020

Country	*Corruption*	*Governance*	*Political Stability*	*Regulatory Quality*	*Rule of Law*	*Voice and Accountability*	*Average Index*
FI	2.20	1.94	1.01	1.86	2.07	1.61	1.78
DK	2.27	1.88	0.95	1.80	1.85	1.52	1.71
LU	2.06	1.83	1.23	1.85	1.78	1.50	1.71
SW	2.13	1.71	1.03	1.69	1.80	1.50	1.64
NL	2.03	1.84	0.86	1.76	1.75	1.52	1.63
AT	1.51	1.65	0.91	1.41	1.80	1.40	1.45
DE	1.86	1.35	0.68	1.59	1.55	1.38	1.40
IE	1.56	1.47	0.98	1.48	1.49	1.39	1.40
EE	1.61	1.33	0.72	1.54	1.37	1.17	1.29
BE	1.48	1.12	0.54	1.35	1.36	1.28	1.19
FR	1.15	1.24	0.32	1.20	1.32	1.07	1.05
PT	0.75	1.02	1.02	0.83	1.18	1.26	1.01
EU27	**0.98**	**1.02**	**0.71**	**1.12**	**1.06**	**1.07**	**0.99**
LT	0.80	1.05	0.93	1.09	0.99	1.01	0.98
CZ	0.58	0.95	0.92	1.24	1.05	0.98	0.96
MT	0.37	1.03	1.01	1.22	0.91	1.12	0.94
SI	0.80	1.16	0.71	0.92	1.06	0.94	0.93
LV	0.72	0.87	0.47	1.19	0.95	0.87	0.85
ES	0.74	0.89	0.44	0.77	0.89	1.01	0.79
CY	0.37	0.88	0.31	1.00	0.57	0.91	0.67
SK	0.44	0.54	0.63	0.79	0.67	0.88	0.66
PL	0.65	0.36	0.52	0.86	0.53	0.62	0.59
IT	0.54	0.39	0.43	0.50	0.24	1.06	0.53
HU	0.09	0.57	0.84	0.48	0.51	0.39	0.48
HR	0.20	0.46	0.61	0.37	0.26	0.58	0.41
EL	0.06	0.44	0.13	0.55	0.32	0.97	0.41
RO	-0.07	-0.26	0.53	0.36	0.39	0.59	0.26
BG	-0.30	-0.18	0.42	0.47	-0.11	0.26	0.09

Source: **Own elaboration based on World Bank data (Worldwide Governance Indicators)**
Note: Some of the variable names were shortened to save space. Corruption measure stands for "Control of Corruption: Estimate," Governance for "Government Effectiveness: Estimate," and Political Stability for "Political Stability and Absence of Violence/Terrorism: Estimate." The average index was simply calculated as an unweighted average of the six institutional variables. All six variables are aggregate indicators, in units of a standard normal distribution, ranging from −2.5 to +2.5

The gap in the quality of institutions and governance between northern and western vs. the southern and eastern EU countries is well established and seems to create a long-lasting cleavage in the deepening European integration (e.g., Farkas, 2019). Thus, it comes as no surprise that the countries with lower-quality institutions evince lower levels of economic development and vice versa (reverse causality), which subsequently also spills over into the lower levels of capacity to be resilient (see also Fig. 4.5). Figure 4.13 clearly shows a positively correlated trend between the institutional quality and the resilience capacity across the EU countries should not be a surprising observation at this point given the endogeneity in institutions' quality and economic development.

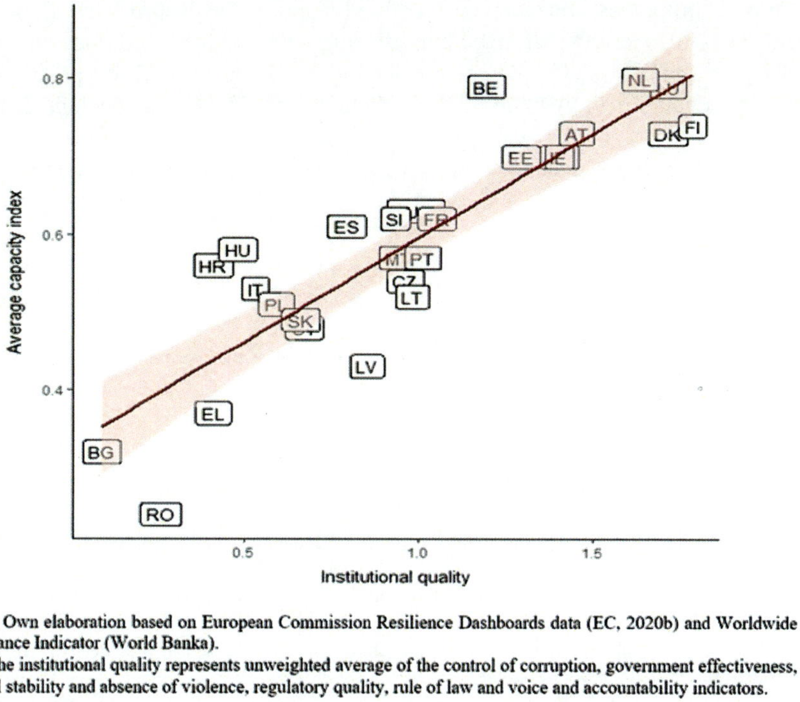

Fig. 4.13 Association between institutional quality and resilience capacity, EU countries 2020. *Note*: The institutional quality represents an unweighted average of the control of corruption, government effectiveness, political stability, absence of violence, regulatory quality, rule of law, and voice and accountability indicators. (*Source*: Own elaboration based on European Commission Resilience Dashboards data (EC, 2020b) and Worldwide Governance Indicator (World Bank))

Contextual factors, namely the quality of institutions and governance, represent additional determinants of the distribution of resilience across the EU countries. A clear divide in economic capacity and institutional endowment between the core (North and West) and peripheral (East and South) EU countries mirrors itself in the capacity to be resilient against nonlinear disturbances. The question remains whether the contextual factors can also be applied to the overall governance of social tipping points within the social–ecological system. In our opinion, the answer to this question is undeniably positive. Even Cutter et al. (2010) argue that disaster resilience can be proxied via a combination of variables gauging social, economic, *institutional*, and infrastructural resilience. And there are numerous research projects that would support this statement. For instance, conducting research on transboundary basins, De Stefano et al. (2012) demonstrate that institutional capacity building is not some abstract issue beyond the boundaries of the ecological systems. Conducting research on transboundary river basins, the authors argue that institutional capacity building is resilient in the fight against climate change. Grefalda et al. (2020) come to a similar conclusion in the context of local non-state actors in

Aurora, Philippines, and their capacity for climate change disturbances. Garschagen (2013) is also critical of underemphasizing institutional conditions in climate change adaptation and resilience conducting empirical research on urban climate change adaptation in Vietnam. Thus, without any doubt, we can conclude that institutional capacity and the quality of governance are essential determinants of governance and resilience even beyond the boundaries of socioeconomic systems.

References

Adger, W. N. (2000). Social and ecological resilience: Are they related? *Progress in Human Geography, 24*(3), 347–364.

Bohle, D., & Greskovits, B. (2012). *Capitalist diversity on Europe's periphery*. Cornell University Press.

Bohle, D., & Regan, A. (2022). Global capital and national growth models: The cases of Ireland and Latvia. In L. Baccaro, M. Blyth, & J. Pontusson (Eds.), *Diminishing returns: The new politics of growth and stagnation* (pp. 323–347). Oxford University Press.

Boschma, R. (2015). Towards an evolutionary perspective on regional resilience. *Regional Studies, 49*(5), 733–751.

Bristow, G., & Healy, A. (2014). Regional resilience: An agency perspective. In *Regional Studies, 48*(5), 923–935.

Bristow, G., & Healy, A. (2018). Innovation and regional economic resilience: An exploratory analysis. *The Annals of Regional Science, 60*(2), 265–284.

Bristow, G. et al. (2014). Economic crisis: Resilience of regions. Draft scientific report. Applied Research 2013/124/2012. In *Luxembourg: ESPON*. Available at https://www.espon.eu/sites/default/files/attachments/Scientific_report.pdf.

Caraveli, H. (2017). The dynamics of the EU core-periphery division: Eastern vs. southern periphery – A comparative analysis from a new economic geography perspective. In G. C. Pascariu & M. A. P. da Silvia Duarte (Eds.), *Core-periphery patterns across the European Union: Case studies and lessons from Eastern and Southern Europe* (pp. 3–22). Emerald Publishing.

Cheng, T., Zhao, Y., & Yhao, C. (2022). Exploring the spatio-temporal evolution of economic resilience of economic resilience in Chinese cities during the COVID-19 crisis. *Sustainable Cities and Society, 84*, 103997.

Christoforou, A. (2010). Social capital and human development an empirical investigation across European Countries. *Journal of Institutional Economics, 6*(2), 191–214.

CLES. (2010). Productive local economies: Creating resilient places. In *Centre for Local Economic Strategies*. Available at https://cles.org.uk/wp-content/uploads/2016/10/Resilience-for-web1.pdf.

Crespo, J., Suire, R., & Vicente, J. (2013). Lock-in or lock-out? How structural properties of knowledge networks affect regional resilience. *Journal of Economic Geography, 14*(1), 199–219.

Cuadrado-Roura, J. R., & Maroto, A. (2016). Unbalanced regional resilience to the economic crisis in Spain: A tale of specialization and productivity. *Cambridge Journal of Regions, Economy and Society, 9*(1), 153–178.

Cutter, S., Burton, C., & Emrich, C. (2010). Disaster resilience indicators for benchmarking baseline conditions. *Journal of Homeland Security and Emergency Management, 7*(1), 1–22.

Cvetkovich, G., & Löfstedt, R. E. (Eds.). (2013). *Social trust and the management of risk*. Earthscan, Routledge.

De Stefano, L. et al. (2012). Climate change and the institutional resilience of international river basins. In *Journal of Peace Research, 49*(1),193:209.

Davies, A., & Tonts, M. (2010). Economic diversity and regional socio-economic Performance. *Geographical Research, 48*, 223–234.
Di Caro, P. (2017). Testing and explaining economic resilience with an application to Italian regions. *Papers in Regional Science, 96*(1), 93–113.
Di Caro, P., & Fratesi, U. (2018). Regional determinants of economic resilience. *The Annals of Regional Science, 60*, 235–240.
Dincecco, M., & Katz, G. (2014). State capacity and long-run economic performance. *The Economic Journal, 126*, 189–218.
Diodato, D., & Weterings, A. B. (2014). The resilience of regional labor markets to economic shocks: Exploring the role of interactions among firms and workers. *Journal of Economic Geography, 15*(4), 723–742.
EC. (2019). Communication from the commission: The European green deal. In *European Commission*. Available at https://eur-lex.europa.eu/legal-content/EN/TXT/?qid=1576150542719&uri=COM%3A2019%3A640%3AFIN.
EC. (2020a). Resilience dashboards. In *European Commission*. Available at https://commission.europa.eu/strategy-and-policy/strategic-planning/strategic-foresight/2020-strategic-foresight-report/resilience-dashboards_en.
EC. (2020b). Europe's moment: Repair and prepare for the next generation. In *European Commission*. Available at https://eur-lex.europa.eu/legal-content/EN/TXT/?uri=COM%3A2020%3A456%3AFIN.
EC. (2020c). 2020 strategic foresight report: Charting the course towards a more resilience Europe. In *European Commission*. Available at https://commission.europa.eu/strategy-and-policy/strategic-planning/strategic-foresight/2020-strategic-foresight-report_en#documents.
EC. (2021a). 2030 digital compass: The European way for the digital decade. In *European Commission*. Available at https://commission.europa.eu/strategy-and-policy/priorities-2019-2024/europe-fit-digital-age/europes-digital-decade-digital-targets-2030_en.
EC. (2021b). Resilience dashboards annex. In *European Commission*. Available at https://commission.europa.eu/document/download/86e24c51-4ec4-4500-a62b-3c57a4f0f6dc_en?filename=dashboard_report_annex_20211129_en.pdf.
EC. (2022). 2022 country report – Luxembourg. In *European Commission*. Available at https://commission.europa.eu/document/download/3c93bb73-e2c6-4972-a797-5024e64f88c9_en?filename=2022-european-semester-country-report-luxembourg_en.pdf.
Egan, M., & Guimarães, M. H. (2017). The single market: Trade barriers and trade remedies. *JCMS: Journal of Common Market Studies, 55*(2), 294–311.
Eurostat. (2019). Eurostat regional yearbook, 2019 edition. In *Eurostat*. Available at https://ec.europa.eu/eurostat/documents/15234730/15245478/KS-HA-19-001-EN-N.pdf/16630b58-99cd-e7b8-9f24-ed8c1227f3c2?t=1667251793754.
Evans, R., & Karecha, J. (2014). Staying on top: Why is Munich so resilient and successful? *European Planning Studies, 22*(6), 1259–1279.
Ezcurra, R., & Rios, V. (2019). Quality of government and regional resilience in the European Union. Evidence from the Great Recession. In. *Papers in Regional Science, 98*(3), 1267–1290.
Faggian, A., et al. (2018). Regional economic resilience: The experience of the Italian local labor systems. *The Annals of Regional Science, 60*, 393–410.
Farkas, B. (2019). Quality of governance and varieties of capitalism in the European Union: Core and periphery division. *Post-Communist Economies, 31*(5), 563–578.
Filippetti, A., et al. (2020). Are innovative regions more resilient? Evidence from Europe in 2008-2016. *Economia Politica, 37*, 807–832.
Fratesi, U., & Rodríguez-Pose, A. (2016). The crisis and regional employment in Europe: What role for sheltered economies? *Cambridge Journal of Regions, Economy and Society, 9*(1), 33–57.
Garschagen, M. (2013). Resilience and organizational institutionalism from a cross-cultural perspective: An exploration based on urban climate change adaptation in Vietnam. *Natural Hazards, 67*, 25–46.

Giannakis, E., & Bruggeman, A. (2015). Economic crisis and regional resilience: Evidence from Greece. *Papers in Regional Science, 96*(3), 451–476.

Giannakis, E., & Bruggeman, A. (2017). Determinants of regional resilience to economic crisis: A European perspective. *European Planning Studies, 25*(8), 1394–1415.

Grabner, S. M. (2021). Regional economic resilience: Review and outlook. In R. Wink (Ed.), *Economic resilience in regions and organizations* (pp. 21–55). Springer.

Graeff, P., & Svendsen, G. T. (2013). Trust and corruption: The influence of positive and negative social capital on the economic development in the European Union. *Quality & Quantity, 47*, 2829–2846.

Grefalda, L. B., et al. (2020). Building institutional resilience in the context of climate change in Aurora, Philippines. *Environmental Research, 186*.

Groot, S. P. T., et al. (2011). The crisis sensitivity of European countries and regions: Stylized facts and spatial heterogeneity. *Cambridge Journals of Regions, Economy and Society, 4*(3), 437–456.

Gür, N. (2020). Does social trust promote behavior aimed at mitigating climate change? *Economic Affairs, 40*(1), 36–49.

Hadjimichalis, C., & Hudson, R. (2014). Contemporary crisis across Europe and the crisis in regional development theories. *Regional Studies, 48*(1), 208–218.

Han, Y., & Goetz, S. J. (2019). Predicting US country economic resilience from industry input-output accounts. *Applied Economics, 51*(19), 2019–2028.

Hassink, R. (2010). Regional resilience: A promising concept to explain differences in regional economic adaptability? *Cambridge Journal of Regions, Economy and Society, 3*(1), 45–48.

Healy, A., & Bristow, G. (2019). Borderlines: Economic resilience on the European Union' eastern periphery. In G. Rouet & G. C. Pascariu (Eds.), *Resilience and the EU's eastern neighborhood countries: From theoretical concepts to a normative agency*. Palgrave Macmillan.

Hu, Y., Li, L., & Dong, K. (2022). What matters for regional economic resilience amid COVID-19? Evidence from cities in Northeast China. *Cities, 120*, 1–11.

IPCC. (2022a). *Climate change 2022: Impacts, adaptation and vulnerability*. Working Group II Contribution to the IPCC Sixth Assessment Report of the Intergovernmental Panel on Climate Change. Available at https://www.ipcc.ch/report/ar6/wg2/downloads/report/IPCC_AR6_WGII_FullReport.pdf.

IPCC. (2022b). *Climate change 2022: Mitigation of climate change*. Working Group III Contribution to the IPCC Sixth Assessment Report of the Intergovernmental Panel on Climate Change. Available at https://www.ipcc.ch/report/ar6/wg3/downloads/report/IPCC_AR6_WGIII_Full_Report.pdf.

Johnston, A., & Matthijs, M. (2022). The political economy of the Eurozone's post-crisis growth model. In L. Baccaro, M. Blyth, & J. Pontusson (Eds.), *Diminishing returns: The new politics of growth and stagnation* (pp. 117–142). Oxford University Press.

Kitsos, A., & Bishop, P. (2018). Economic resilience in Great Britain: The crisis impact and its determining factors for local authority districts. *The Annals of Regional Science, 60*, 329–347.

Martin, R., & Sunley, P. (2015). On the notion of regional economic resilience: Conceptualization and explanation. *Journal of Economic Geography, 15*(1), 1–42.

Martin, R., et al. (2016). How regions react to recessions: Resilience and the role of economic structure. *Regional Studies, 50*(4), 534–585.

Matthijs, M., & Blyth, M. (2015). *The future of the Euro*. Oxford University Press.

Matthijs, M., & McNamara, K. (2015). The euro crisis' theory effect: Northern saints, southern sinners, and the demise of the Eurobond. *Journal of European Integration, 37*(1), 229–245.

Milkoreit, M., et al. (2018). Defining tipping points for social-ecological systems scholarship – An interdisciplinary literature review. *Environmental Research Letters, 13*, 1–13.

Modica, M., & Reggiani, A. (2015). Spatial economic resilience: Overview and perspective. *Networks and Spatial Economics, 15*, 211–233.

Murua, J. R., & Ferrero, A. M. (2019). Talking about regional resilience: Evidence from two formerly rural Spanish regions. *European Planning Studies, 27*(11), 2312–2328.

Neff, P. (1949). Interregional cyclical differentials: Causes, measurement and significance. *American Economic Review, 39*, 105–119.

Nölke, A., & Vliegenthart, A. (2009). Enlarging the varieties of capitalism: The emergence of dependent market economies in east central Europe. *World Politics, 61*(4), 670–702.

North, D. C. (1990). *Institutions, institutional change and economic performance*. Cambridge University Press.

Obschonka, M., et al. (2016). Macropsychological factors predict regional economic resilience during a major economic crisis. *Social Psychological and Personality Science, 7*(2). https://doi.org/10.1177/1948550615608402

OECD. (2016). OECD regions at a glance 2016. In *OECD*. Available at https://www.oecd-ilibrary.org/governance/oecd-regions-at-a-glance-2016_reg_glance-2016-en.

Östh, J., Reggiani, A., & Galliazzo, G. (2015). Spatial economic resilience and accessibility: A joint perspective. *Computers, Environment and Urban Systems, 49*, 148–159.

Östh, J., Dolciotti, M., Reggiani, A., & Nijkamp, P. (2018). Social capital, resilience and accessibility in urban systems: A study on Sweden. *Networks and Spatial Economics, 18*, 313–336.

Pontarollo, N., & Serpieri, C. (2020a). A composite policy tool to measure territorial resilience capacity. *Socio-Economic Planning Sciences, 70*, 100669.

Pontarollo, N., & Serpieri, C. (2020b). Towards regional renewal: A multilevel perspective for the EU. *Regional Studies, 54*(6), 754–764.

Prokkola, E. K. (2019). Border-regional resilience in EU internal and external border areas in Finland. *European Planning Studies, 27*(8), 1587–1606.

Puškárová, P. (2022). Trust or bust: Growth effects of knowledge, human and social capital revisited. In *Economic Systems, 46*(4), 1–9.

Rapacki, R., et al. (2019). *Diversity of patchwork capitalism in central and eastern Europe*. Routledge.

Reeves, A., et al. (2015). Economic shocks, resilience, and male suicides in the Great Recession: Cross-national analysis of 20 EU countries. *European Journal of Public Health, 25*(3), 404–409.

Rodríguez-Posé, A. (2013). Do institutions matter for regional development? *Regional Studies, 47*(7), 1034–1047.

Rodríguez-Posé, A., & von Berlepsch, V. (2014). Social capital and individual happiness in Europe. *Journal of Happiness Studies, 15*, 357–386.

Sabatino, M. (2019). Economic resilience and social capital of the Italian region. *International Review of Economics and Finance, 61*, 355–367.

Sánchez-Zamora, P., Gallardo-Cobos, R., & Ceña-Delgado, F. (2014). Rural areas face the economic crisis: Analyzing the determinants of successful territorial dynamics. *Journal of Rural Studies, 35*, 11–25.

Sensier, M., Bristow, G., & Healy, A. (2016). Measuring regional economic resilience across Europe: Operationalizing a complex concept. *Spatial Economic Analysis, 11*(2), 128–151.

Simmie, J., & Martin, R. (2010). The economic resilience of regions: Towards an evolutionary approach. *Cambridge Journal of Regions, Economy and Society, 3*, 27–43.

Storper, M., & Scott, A. (2009). Rethinking human capital, creativity and urban growth. *Journal of Economic Geography, 9*, 147–167.

Sutton, J., & Arku, G. (2022). Regional economic resilience: Towards a system approach. *Regional Studies, Regional Science, 9*(1), 497–512.

Sutton, J., et al. (2023). Regional economic resilience: A scoping review. *Progress in Human Geography, 47*(2), 030913252311741.

Tennekes, M. et al. (2022). Tmap: Thematic maps. In *CRAN*. Available at https://cran.r-project.org/web/packages/tmap/tmap.pdf.

Vining, R. (1946). Location of industry and regional patterns of business cycle behavior. *American Economic Review, 39*, 89–104.

Wang, Z., & Wei, W. (2019). Regional economic resilience in China: Measurement and determinants. *Regional Studies, 55*(7), 1228–1239.

World Economic Forum (2020). The global competitiveness report. How countries are performing on the road to recovery. In *World Economic Forum.* Available at https://www.weforum.org/reports/the-global-competitiveness-report-2020/.

Wu, D., et al. (2023). An agency perspective of regional economic resilience during COVID-19: The role of the local state's place-based leadership in Kunshan, China. *Cambridge Journal of Regions, Economy and Society, 16*(1), 119–133.

Xiao, J., Boschma, R., & Andersson, M. (2018). Resilience in the European Union: The effect of the 2008 crisis on the ability of regions in Europe to develop new industrial specialization. *Industrial and Corporate Change, 27*(1), 15–47.

Chapter 5
Governing Social Tipping Points in the EU's Periphery: A Conceptual Framework and Methodology

So far in this monograph, we have conceptualized the concept of social tipping points within social–ecological systems and possibilities of multilevel agency within the European Union (EU) context to have the capacity to govern and be resilient against such abrupt, nonlinear perturbations having multiple social–ecological side effects. We went through theoretical concepts of governance, state capacity, state resilience, and non-state actors trying to put an emphasis on the stakeholders' agency when governing nonlinear disturbances. Furthermore, we also observe that the capacity to govern and be resilient against these perturbations within the EU context is also spatially and geographically determined and relies on compositional, collective, and contextual factors. In this section, we follow up on these theoretical concepts and provide a conceptual framework for the governance of social tipping points with a special focus on the EU's periphery. Then, we outline our empirical strategy and methodology to test whether this framework holds against real cases.

5.1 Governance of Social Tipping Points: A Conceptual Framework

Let us first summarize all the concepts we have gone through so far in this book, intending to conceptualize the inter-dynamics between governance, resilience, and social tipping points and their position with regard to different actors. We intend to draft a conceptual framework for the governance of tipping points and the role of the state and non-state actors in it (Fig. 5.1). Returning to Chap. 2, we follow up Milkoreit et al.'s (2018) conceptualization of *social tipping points* as relatively small qualitative changes within a social–ecological system that trigger an abrupt, nonlinear change in the social component of that system, are driven by a self-reinforcing positive feedback mechanism, and by this account inevitably lead to a qualitatively different state of the social–ecological system with often limited reversibility (hysteresis). These four assumptions – multiple states, abruptness, feedback, and limited irreversibility – form the backbone of our definition of a social tipping point. Every social tipping point has numerous amplifiers and triggers, either external or internal, be it migration, geopolitics, or climate (Szabó & Jančovič, 2020). All these amplifiers constantly add to the likelihood of the

J. Szabó et al., *Governance of Social Tipping Points*, Key Challenges in Geography,
https://doi.org/10.1007/978-3-031-47413-2_5

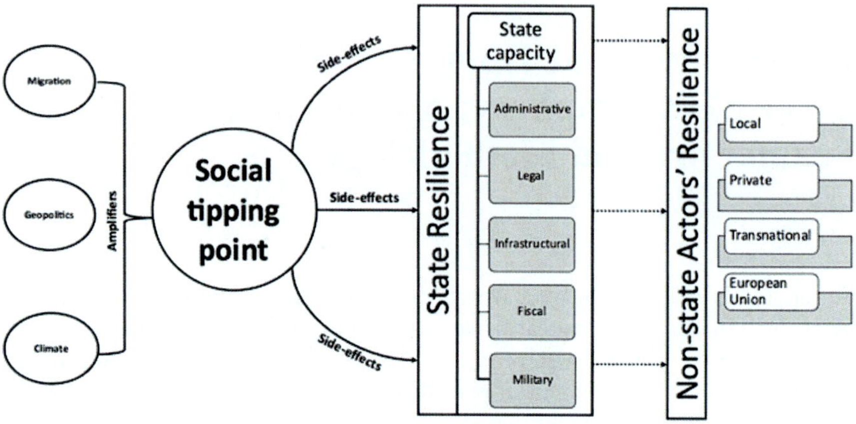

Source: Own elaboration.

Fig. 5.1 Governance of social tipping points: a conceptual framework. (*Source*: Own elaboration)

emergence of a social tipping point. Given the unexpectedness and abruptness of social tipping points, no authority can credibly anticipate the occurrence of social tipping points and their side effects, such as increased inflows of migrants and refugees, pest epidemics, or collapse of a local economy, and is constrained by the lack of information about the future nonlinear events. The social tipping points and especially their side effects stemming from the abrupt perturbation might be governed though, depending on the capacity and resilience of the crucial stakeholders.

Earlier in Sect. 3, it was established that governance represents a dynamic process of the general exercise of authority, where authority represents first and foremost state institutions that select goals, subsequently reconcile, and coordinate these goals among various factions within the state to implement the most acceptable goal given the material conditions. Then, governance ends up with feedback and accountability, especially in the context of the (European) shackled Leviathan (Acemoglu & Robinson, 2020). We adopted a functionalist and openly state-centered concept of governance since we are of the opinion the state remains at the center of governance of social tipping points, in spite of the recent trends in globalization or, on the contrary, decentralization, mainly due to the *Kompetenz-Kompetenz* and essential monopoly of the legitimate use of the force on its territory. The main determinant of the governance of the social tipping point is the state capacity to implement its policies (*state capacity*) and the capacity to withstand disturbances stemming from the social tipping point and alternatively adapt to them (*resilience*). Since the state capacity can be viewed as the ability of the state to implement its defined goals and depends on the extent of expected public goods the state ought to and can deliver, it consists of five subcomponents (Table 3.1). These subcomponents are administrative, legal, infrastructural, fiscal, and defense capacities and in total determine the state capacity. The administrative capacity ensures the day-to-day administration of the state, the legal capacity covers the rule of law and protection of human rights and liberties, the infrastructural capacity administers and sustains various aspects of the state's critical infrastructure, the fiscal capacity makes sure enough resources are at disposal for the state and these are properly redistributed, and finally, the military capacity determines the overall capacity of all law-enforcement units (e.g., police, army, etc.). Furthermore, it was argued that state capacity and resilience are two faces of the same coin. State capacity, or the state's ability to implement its defined policies, determines the state's resilience, or the amount of disturbance that a system can undergo while retaining its previous functions and at the same time the extent to which a system is able to reorganize and adapt. Resilience thus represents the persistence and ability to withstand disturbances along with the capacity to recover and adapt.

Figure 5.1 summarizes the inter-dynamics between the concepts of social tipping points, governance, state capacity, and state resilience. Given our already mentioned state-centric approach, it comes as no surprise that the state, determined by its state capacity, stands at the forefront of society, facing the impacts of social tipping points along with side effects bombarding the state's resilience. The greater the state capacity in the administrative, legal, infrastructural, fiscal, and military realms, the more resilient the state is in withstanding the disturbances and capable of reorganizing and adapting if need be. However, there often comes a time when the state capacity and resilience just fall short of the overall impact stemming from social tipping points and their impacts soak through the state, which functions (should function) as a protective umbrella over its polity. Even if the state intercepts most of the social tipping points' impacts, there are always some disturbances that can soak through and impact the society (ergo dashed arrows). This is where the multilevel governance comes at hand (Fig. 3.4). The capacity (the executive branch of governance) spans multiple levels going beyond the state, which is, naturally, at the center. Then, the society has also at its disposal other non-state actors taking part in governance at various levels – transnational or subnational – that are increasingly able to shape the conduct of other actors, but primarily the state. National actors, both local (regional local authorities, local NGOs, and local civil society) and private (private companies, nonprofit organizations (NPOs), and NGOs), along with transnational actors (international organizations (IOs) and private transnational actors) and the EU dispose over their respective capacities that are mirrored in their respective abilities to withstand nonlinear perturbations and adapt to them. The EU as a transnational actor occupies a unique position in the universe of multilevel governance within the context of the EU member states due to its formative impact on the constitutional aspects of its member states. Since these non-state actors constitute part of the polity (society), they tend to mobilize and complement the state in governing social tipping point and its side effects. Non-state actors are interested in mitigating the impacts of social tipping point since it tends to decrease the overall society's welfare, and thus secondarily subsidize the state. That is why even non-state actors invest their resources to support the state, although the state was first and foremost created as Hobbesian Leviathan to avoid the "Warre," including social tipping points (Acemoglu & Robinson, 2020). In our conceptualization, this is why the non-state actor's resilience is, consisting of the sum of fractional resiliencies of the non-state actors that are, in turn, determined by their respective capacities, important. The sum of societal resilience is therefore determined by the state and non-state resilience and their capacities. Ultimately, it is the individual and collective agency at different spatial scales that shape how societies rebound from shocks (Martin & Sunley, 2015). How does the state benefit from the non-state actors? One of the main arguments of this monograph is that it is in the state's greatest interest to invest in the capacity and resilience building of the non-state actors since such investments yield a return on the overall societal resilience, the state included. If the overall societal resilience is insufficient vis-à-vis the impacts stemming from a social tipping point, the fragility of the state augments (or reversely the strength of the state declines), which makes the state greatly dependent on the capacity and resilience of the non-state actors. "The key to resilience is thus adaptability, which is enabled by the nonlinear nature of the relationship between constituent parts of the system" (Fjäder, 2014: 120).

The last component of our argument goes beyond the stakeholder's agency as a prerequisite for social tipping points governance. Since our focus in this monograph is mostly directed at the peripheral EU countries, it is also important to take into account spatial and geographical determinants of tipping points governance, capacity, and resilience. In Sect. 4, it was demonstrated that there is a discrepancy between how different EU regions, namely the core–periphery dichotomy, vary in their capacity and resilience vis-à-vis abrupt, nonlinear perturbations within social–ecological systems that have self-reinforcing feedback effects. Furthermore, the spatial distribution of vulnerabilities and capacities with regard to resilience across the EU member states is seeming to the detriment of the eastern (Visagrad countries – Czechia, Hungary, Poland, and Slovakia), southeastern (Bulgaria and Romania), and southern (Cyprus, Greece, Spain, Portugal, and Italy) periphery (Figs. 4.1, 4.2, 4.3, and 4.4 as well as Table 4.1). Furthermore, a clear association between the EU member states' economic development proxied via national income and their resilience capacity was demonstrated giving us cogent enough evidence that the ability to withstand nonlinear disturbances is positively conditioned upon the level of economic development (Fig. 4.5). As a result, this allows us to come to a relatively counterintuitive argument that not only do the member states need to invest in the non-state actors' capacity and resilience, including the ones belonging to the European Union, but the European Union and the core member states will also benefit from strengthening the Union's peripheral countries and regions in order to make the European Union and themselves more resilient against social tipping points. Furthermore, drawing from the field of regional economic resilience, it was argued that there are also additional determinants influencing the spatial distribution of resilience – compositional, collective, and contextual (Fig. 4.6) – that can be applied to the geographical distribution of resilience across the EU. Compositional factors are closely related to structural composition and collective factors refer to networks and dependency beyond the stakeholders, whereas contextual factors describe the importance of institutions. All these factors confirm the abovementioned peripheral EU regions evince the lowest levels of resilience against social tipping points.

5.2 Methodology and Empirical Strategy

In the preceding section, a deductive-nomological model was constructed (also covering law model), conceptualizing how governance, capacity, and resilience mutually interact when facing social tipping points and side effects stemming from them at multilevel governance of state and non-state actors. In this section, we aim to outline our methodology and empirical strategy to determine whether the conceptual framework developed in Sect. 5.1 holds against real-life cases. In a nutshell, we are interested in decoding the following research question (RQ): "*What factors determine the success or failure of social tipping point governance in the EU's eastern and southern Periphery*?" Based on the RQ that follows up our conceptual framework constrained by spatial determinants, two hypotheses can be laid down:

(a) *"Successful governance of the social tipping points depends on well-developed state capacity and state resilience."*
(b) *"If the state capacity and state resilience fall short of social tipping point and its side effects and some of them soak through the protective umbrella of the state, non-state actors (local, private and transnational) tend to complement the state with their respective capacities and resilience and avert the state from the decline in its strength via withstanding the non-linear disturbance or adopting to it."*

In order to be able to verify or falsify our compound hypothesis, we have to (1) *identify the presence of a social tipping point*, (2) *determine how the state proved itself when facing this particular social tipping point and how resilient it was,* and (3) *to determine the role of the non-state actors in dealing with that social tipping point.* Since our conceptual framework (deductive-nomological model) needs to be verified with empirical observations, we opt for a multiple-case study research design that will allow us to capture multiple intersectional realities that are not easily quantifiable. Case study approach to study resilience is relatively widespread in the academia (e.g., Simmie & Martin, 2010; Cowell, 2013; Evans & Karecha, 2014). Thus, we examine three case studies, each of them depicting a potential social tipping point in a certain EU peripheral country, however, each with a different trigger. This way, more robust and valid results will be secured. Three case studies were selected for (theoretical) replication logic (Yin, 2014). The first case study aims to identify the migration-induced social tipping point, the second climate-induced social tipping point, and the third geopolitics-induced social tipping point. Migration was chosen due to the social aspect, climate due to the environmental aspect, and geopolitics due to the international aspect. Furthermore, each case study was positioned in a different EU peripheral country, one in the south (Greece) and two in the east (Slovakia and Poland). Diverse country-level research designs for assessing governance seem to remain the most suitable research designs (Sagan et al., 2021).

Given the scope of this monograph, we focus on three social tipping points mostly impacting countries in the EU's eastern and southern borders as shown in

Sect. 4, they are most susceptible to exogenous shocks due to low capacity and resilience. As a potential migration-induced social tipping point, we take a look at the most recent migration and refugee crisis hitting Europe in 2015, with a focus on the subsequent consequences stemming from this crisis. We will especially focus on the strain on Moria refugees on the Greek island of Lesvos that resulted in the camp burnout in September 2020. As a potential climate-induced social tipping point, we analyze the often overlooked (mis)management of European forests. Our primary focus will be on the resilience vis-à-vis climate-induced windstorm affecting the Tatra Mountain in Slovakia in 2004, which directly resulted in a pest epidemic, the bark beetle outbreak in Slovak Tatra National Park. As a geopolitics-induced social tipping point, we cannot ignore the recent shifts East of the European border – the Russian aggression in Ukraine. This potential social tipping point has multiple negative side effects, among which the massive exodus of Ukrainians into Europe and gas spike prices are prominent.

The subsequent section thus contains three subsections, each one containing individual case reports for each case study. Each case report is also divided into three parts, where each part aims to contribute to verifying or falsifying our hypotheses via:

- *Identifying whether that case represents a social tipping point*
- *Analyzing how the state proved itself when facing this particular social tipping point and how resilient it was*
- *Determining the role of the non-state actors in dealing with that social tipping point*

Based on cross-case conclusions stemming from our case study reports, we modify our theory and develop policy recommendations in the last section. This will be done through deductive theory conceptualization based on data from within each case (Johanssonn, 2007). It is attempted to show that these three particular cases can be synthesized and help us to verify our hypothesis and to reflect our conceptual framework from the previous chapter. Such a research design will allow us to secure validity (construct, internal, and external) as well as reliability of our research (Yin, 2014).

As for our data, we collect mostly qualitative, but also quantitative supplementary materials that can help us to meet our stated aims, verify or falsify the hypotheses, and answer the RQ. In terms of collecting the data, we always proceed in the following order. We first collect and study all research papers at least partially touching upon the analyzed case. The research papers are looked for using the Google Scholar search engine. We look up all relevant research papers by typing in the keywords. Second, once all scientific papers are analyzed, we proceed to analyze policy briefs, working papers, and documents issued by key stakeholders, namely NGOs, IO, institutions of the EU (namely the Commission), and the state authorities. If no sufficient and convincing information was found there, we proceeded to search newspaper articles using the Google Search engine. Only relevant and internationally acclaimed newspapers and online portals were analyzed. If such could not be found, we also analyzed local newspapers. Whenever possible, we also try to supplement the qualitative data with quantitative data.

References

Acemoglu, D., & Robinson, J. A. (2020). *The narrow corridor: States, societies, and the fate of liberty* (2nd ed.). Penguin Press.

Cowell, M. (2013). Bounce back or move on: regional resilience and economic development planning. *Cities, 30*, 212–222.

Evans, R., & Karecha, J. (2014). Staying on top: Why is Munich so resilient and successful? *European Planning Studies, 22*, 1259–1279.

Fjäder, C. (2014). The nation-state, national security and resilience in the age of globalization. *Resilience, 2*(2), 114–129.

Johanssonn, R. (2007). On case study methodology. *Open House International, 32*(3), 48–54.

Martin, R., & Sunley, P. (2015). On the notion of regional economic resilience: Conceptualization and explanation. *Journal of Economic Geography, 15*(1), 1–42.

Milkoreit, M., et al. (2018). Defining tipping points for social-ecological systems scholarship – An interdisciplinary literature review. *Environmental Research Letters, 13*, 1–13.

Sagan, A., et al. (2021). Health system resilience during the pandemic: It's mostly about governance. *Eurohealth, 27*(1), 10–15.

Simmie, J., & Martin, R. (2010). The economic resilience of regions: Towards an evolutionary approach. *Cambridge Journal of Regions, Economy and Society, 3*, 27–43.

Szabó, J., & Jančovič, P. (2020). Social tipping points & asylum seekers: Fragility in the EU's remote island areas. *In International conference on European integration 2022*, Ostrava, pp. 694–707.

Yin, R. K. (2014). *Case study research: Design and methods* (5th ed.). Sage Publications.

Chapter 6
Resilience in Migration, Climate Change, and Geopolitics: A Case of the EU's Periphery

This chapter summarizes case reports of three potential cases of social tipping points in the European Union (EU) periphery: migration-induced, climate change–induced, and geopolitics-induced abrupt changes triggering adverse consequences in Greece (southern EU periphery) as well as Slovakia and Poland (eastern EU periphery). Each case study begins with a general description of the situation leading to the possible eruption of a social tipping point and tries to identify amplifiers that played a role in triggering the tipping point and potential consequences stemming from it. Subsequently, the capacity and resilience of state authorities of Greece, Slovakia, and Poland are analyzed in order to assess the governing of the respective social tipping point, with the aim of determining how crucial the state capacity and resilience are when it comes to governing social tipping point. Subsequently, given our focus on multilevel governance of social tipping points, the role of the non-state actors, namely transnational, private, and local, is examined as well with the goal of analyzing their dynamics with the state when facing social tipping points and their consequences.

6.1 Moria Camp Burndown

6.1.1 Identification of the Migration-Induced Social Tipping Point

As we already outlined in the introduction, the migration inflows into Europe were preceded by climate-related aspects, namely climate hazards (droughts and floods), that directly and indirectly contributed to the massive exodus of people from the fragile areas impacted by climate change, mainly in Africa and the Middle East and North Africa (MENA) region. As a result, climate change and associated extreme climate events influenced already vulnerable communities via the destruction of shelters or the (im)possibility of earning one's livelihood, causing internal displacements. Facing unbearable climate and food insecurity, numerous violent conflicts erupted as fragile communities left their homelands for urban centers or neighboring African countries and later continued their journeys to continental Europe. One

J. Szabó et al., *Governance of Social Tipping Points*, Key Challenges in Geography,
https://doi.org/10.1007/978-3-031-47413-2_6

of the more substantial conflicts causing rising migration to Europe was the civil war erupting in Syria in March 2011, which was closely associated with the rise of the Islamic State of Iraq and Syria. Subsequently, more than 12 million Syrians have been forced to leave behind their homeland, with a refuge population of 4.2 million by mid-2015, replacing Afghanistan as the main source of refugees worldwide (Crawley, 2016: 15). Although many of these people ended up living in refugee camps in Jordan and Lebanon or urban setting within Turkey, many fled Syria for Europe.

Various European countries had already quite sizeable foreign-born populations prior to 2015. Of Germany's 12 million migrants before the start of the migration crisis in 2015, the largest groups came from Poland, Turkey, the Russian Federation, and Kazakhstan, with each exceeding one million (McAuliffe & Ruhs, 2017). Although the streams of populations from the Middle East and Africa crossing the Mediterranean to reach Europe have been on the rise since 2011, the year 2015 marked the culmination of new arrivals to Europe and deaths in the Mediterranean as the situation in the MENA region drastically deteriorated. One of the most memorable articulations of the peak of the migration crisis was former German Chancellor, Angela Merkel's "Wir schaffen das" speech from August 31, 2015 (Holzberg, 2021). According to the International Organization for Migration (IOM) in 2015 alone, 1,046,599 people migrated to Europe of which 34,887 arrived by land and 1,011,712 by sea (IOM, 2016). This trend continued well into 2016 when more than 300,000,000 people arrived in Europe. Figure 6.1 portrays the immigration flows to Greece as our country of interest in the EU's southern periphery with a clear and sudden peak in 2015. From the social tipping point (STP) perspective, the migration peak into Europe in 2015 represents a threshold causing an abrupt nonlinear change tipping the previous socio-economic system (SES) into a new, qualitatively different state. Once a certain, not precisely defined threshold was crossed, it triggered a

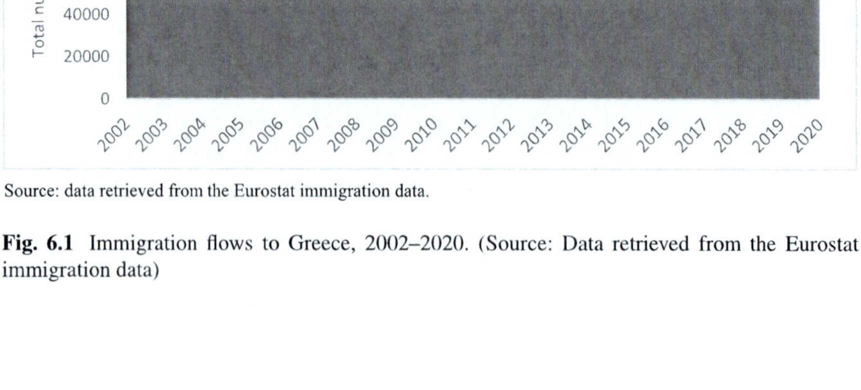

Source: data retrieved from the Eurostat immigration data.

Fig. 6.1 Immigration flows to Greece, 2002–2020. (Source: Data retrieved from the Eurostat immigration data)

self-reinforcing feedback mechanism generating new migration inflows into Europe, with various adverse consequences and side effects. Side effects of such migration-induced STP can be various, ranging from political and economic to social and urban (e.g., Abou-Chadi et al., 2022).

As we can see from Fig. 6.2, the sea route is by far the preferred one. Naturally, the increased migration inflows brought about various adverse consequences coming with the migration by sea, namely human tragedies occurring during the attempts to cross the Mediterranean Sea. A total of 3770 people were recorded dead or missing in 2015, albeit the actual figure is almost certainly higher. According to the Missing Migrants Project (MMP), more than half of the 50,000 deaths in the world documented during migration between 2014 and 2022 occurred within or en route to Europe. Notably, 20,122 deaths and disappearances were recorded on the Central Mediterranean route,[1] 2829 on the Western Mediterranean route,[2] and nearly 2162 on the eastern Mediterranean route,[3] with nearly 803 recorded in 2015 alone (MMP, 2022). This makes Europe easily the deadliest migration destination (Fig. 6.2).

In order to deal with the increasing influx of migrations and unfortunate deaths in the Mediterranean Sea, the European Commission (EC) presented the European Agenda of Migration and announced the development of new hotspots as a way of

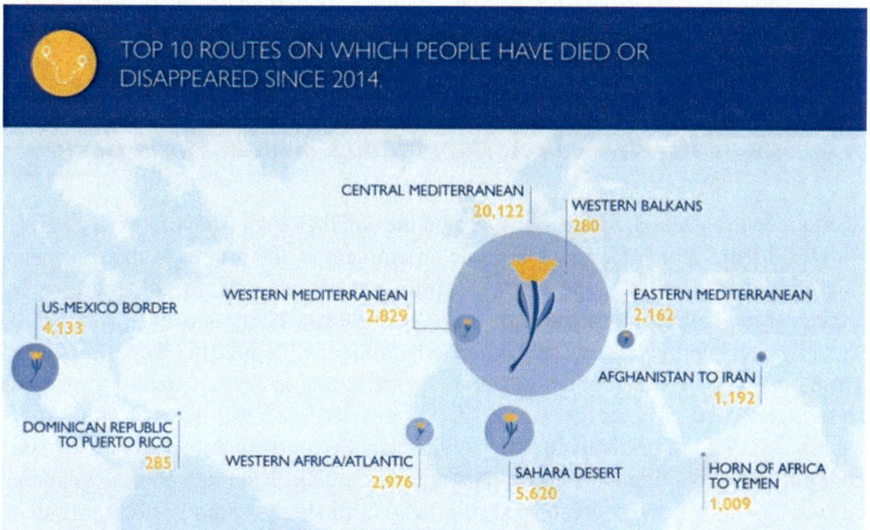

***Source*: Missing Migrants Project (MMP, 2022).**

Fig. 6.2 Top 10 routes on which people have died or disappeared since 2014. (*Source*: Missing Migrants Project (MMP, 2022))

[1] Routes for people crossing the shores of Italy and Malta from sub-Saharan Africa.

[2] Mainly routes to the Spanish mainland near the Spanish enclaves Melilla and Ceuta.

[3] Greece, Cyprus, and Bulgaria.

assisting frontline member states in "managing exceptional migratory flow" (Neville et al., 2016). The hotspot approach stated that the EU agencies (Frontex, European Asylum Support Service, the European Police Office, and the European Union's Judicial Cooperation Unit) should cooperate with member states, namely Italian and Greek authorities, in identification procedures, asylum applications, and return operations within newly established hotspot places, camps near the EU external border. The hotspots were supposed to channel migration flows, letting pass those who are in need of asylum and sending back those who are not (Pollozek & Passoth, 2019). In the following months, 11 new hotspots were established, among which was the Moria Reception and Identification Centre on the Greek island of Lesvos. Ultimately, the Moria hotspot, a place where Europe's ideals, solidarity, and human rights, dissolved in a tangle of bureaucracy, indifference, and violence, ended in tragedy. Moria hotspot on Lesvos island in Greece, which has been described by humanitarians as "the worst refugee camp on earth" or "living hell," was reduced to ashes following the burndown on September 8, 2020 (Gordon & Larsen, 2021). In the subsequent two sections, we take a look at the governance of this migration-induced STP, the European migration crisis peaking in 2015, and the side effects stemming from it, namely the mismanagement and burndown of the Moria reception hotspot. First, we analyze the state response to asylum management in Lesvos and its capacity, and second, we also take a look at non-state actors involved in this tragedy, mainly the EU, NGOs, and local communities.

6.1.2 State Resilience in Asylum Management in Moria Camp

Moria reception center, founded in 2015 on the small Greek island of Lesvos, barely 4 miles off the coast of Turkey, became an emblem of the failure of the European migration policy. Initially, the Moria reception center – a former military base – was a place designed for something over 1200 asylum seekers and even though its capacity was augmented by the Syriza government (2015–2019) to 3200 places, these numbers were still quite deficient. As the strain on Lesvos started to grow – Lesvos recorded 1417 arrivals in 2012, 3233 in 2013, 23,187 in 2014 up until an astonishing 512,327 arrivals in 2015, and Moria became a hub for asylum seekers trying to reach continental Europe (Psaropoulos, 2020). Although Moria's planned capacity was for 3000 refugees and migrants, at times the camp hosted more than 20,000 people (Digidiki & Bhabha, 2020). Due to its inflated capacity, Moria never functioned as a typical detention center but as a camp with monitored and restricted movement – a space of indefinite containment according to Gordon and Larsen (2021). The authors also describe how it was guarded by police and was referred to by temporary residents as *carceral* in nature, resembling a prison because of the perimeter fences, barbed wire, and gates as well as checkpoints and police guards.

The situation in Moria center started deteriorating after the 2016 agreement between the EU and Turkey, preventing numerous asylum seekers from returning to their countries of origin or continuing to mainland Europe. As a result, the number of people "living" in the Moria camp started to build up uncontrollably. Commissioner for Human Rights (CfHR) of the Council of Europe Dunja Mijatović already warned in October 2019 that "the situation of migrants, including asylum seekers, in the Greek Aegean islands, has dramatically worsened" and urgent measures need to be undertaken in order to address the desperate conditions of people living there, especially in Moria (CfHR, 2019). In 2018, UNHCR urged the Greek government to move the Moria asylum seekers away from Lesvos due to reaching a "boiling point." It was at the time when still "only" 8000 people were living in the Moria camp (UNHCR, 2018). Unhygienic conditions, lack of medical care and sanitation, or people waiting in queues for hours just to get to food or to go to the bathrooms were day-to-day norms in Moria. In 2019, a new Greek government was sworn in led by Prime Minister Mitsotakis, who won the parliamentary election partly because of his allegedly tougher stance against migration and asylum seekers compared with his left-wing predecessor, Alexis Tsipras. Although the conservative government initially secured the transfer of over 14,000 people from the island to the mainland between September 2019 and January 2020, another 40,000 new arrivals crossing the borders in the same period counterbalanced these efforts, and it was not enough to reduce the numbers in the hotspot. According to Dimitriadi (2020: 4), "the continuous influx couples with the containment policy on the islands brought local communities to the boiling point." As the presence of migrants in squalid conditions is a relatively new phenomenon for the locals, this also spurred some backlashes at the local level, watching the containment policies and most importantly public infrastructure fail. Subsequently, the government's proposals to construct new detention centers faced vigorous opposition in the form of protests, roadblocks, and legal actions from residents, fearing the repetition of the Moria case. In order to appease locals, an already hamstrung new conservative majority made a promise to the locals to streamline the asylum system to process applications within 28 days, including appeals, rather than the existing average of several months. Naturally, "older" applicants ended up waiting even longer, as the asylum process refocused on achieving its target speed with new arrivals. As the number of asylum seekers started to further accumulate and the asylum process started to get lengthier, frustration grew as well, undermining the already unstable community (Dimitriadi, 2020).

Migrants trapped on the remote island, economically vulnerable and fragile, it is only natural that the list of potential problems accumulates. As the number of asylum seekers exceeded the camp capacity at least fivefold, with over 167 people per toilet, 242 people per shower, and up to 1300 sharing one water tap, not to forget no functioning sewage system, various health problems started to emerge (Gordon & Larsen, 2021). Very poor hygiene (e.g., limited availability of soap) resulted in the spread of numerous dangerous diseases and skin conditions for children. Children suffering from malnutrition had to wait over 3 h long queues for their meals. If that was not enough, medical experts taking care of people in the camp had to deal with suicide attempts of the minors on a daily basis, as registered by Doctors Without Borders (MSF) (MSF, 2018), who were also warning against a potential breaking point. Furthermore, several people suffered from respiratory problems caused by tear gas used by the police to fight the crime in the camp. It comes as no surprise that unemployment and economic frustration caused by this desperate situation automatically bred lawlessness. Toward the end, crime used to be one of the biggest problems in Moria. Murders, rape cases, and drug trafficking became a new normal, with ethnic conflicts erupting into bloody violence day to day. Afghans fighting Arabs, Sunni in constant conflict with Shia, Muslims attacking Kurds, ethnic conflicts, and sectarianism in the camp precisely imitating real geopolitical conflicts prevailed in the Middle East. In addition, German intelligence even identified ISIS branches operating in Moria, creating their own communities based on Sharia law, and no-go zones, taking over large parts of the camp (Pérez-Sales et al., 2022). Even if the Greek authorities eventually did manage to take measures against "delinquency" and detained some people and coerced them to live in the infamous Section B for an uncertain period of time, it almost always ended up counterproductive and repressive (Illiadou, 2019). Peréz-Sales et al. (2022), conducting a cross-sectional study based on the model of torturing environments, argue that the Moria reception centers represent a "space of systemic ill-treatment vulnerating the European legal standards related to torture" (Article 3 of the Human Rights Convention). It is not farfetched to say that, at this point, the Greek state lost control over the Moria camp.

At that time, the problem was that the Greek state lacked the required capacity to deal with all these problems in the Moria center. We must be first aware of the fact that the Greek state had been under scrutiny ever since Greece was severely hit by the Great Financial Crisis in 2007–2008 and afterward suffered economic consequences stemming from it (sovereign crises in 2009, 2010–2012, and 2015). Having been hit hard by one of the worst economic recessions in postwar history, Greece's state capacity, mainly its administrative, legal, and fiscal sections, was significantly weakened. High levels of unemployment in combination with rising numbers of asylum seekers and unrecorded migrant laborers, with no real chance of earning the living, led to social unrest, a rise of petty crime, and protests from the locals (Petrakos & Psycharis, 2016). Subsequently, the Greek state, unable to cope with its socioeconomic and security problems, was forced to abandon the Moria camp. As the crime rate became unsustainable, various charities and NGOs started to leave in protest, criticizing the Greek state's idleness, leaving asylum seekers and locals in an unenviable situation. The only people who helped the whole system to function were local movements and volunteer groups. The system started to get absolutely overwhelmed once the camp started detecting its first positive COVID cases. As doctors and armed forces attempted to isolate the positively tested into special zones, imposing necessary lockdowns (to protect the local population, not camp residents), asylum seekers started rioting, destroying the camp infrastructure, and ultimately setting the whole camp on fire. On the night of September 8 and 9, 2020, the Moria camp with nearly 13,000 asylum seekers (including 4000 unaccompanied minors), supposedly the largest refugee camp in the world, was burnt down to the ground.[4] As a result of this felony, Greek police arrested half a dozen Afghan men

[4] It was not the first time a fire was started in the camp. A few isolated fires resulted in the deaths of asylum seekers living in the camp already before September 2020 (MSF, 2020). Gordon and Larsen (2021) describe how some of their research participants mentioned fires occurring regularly due to technical faults and migrants protesting camp conditions.

on charges of arson, four adults and two unaccompanied minors (Gordon & Larsen, 2021).[5]

The ungoverned migration-induced social tipping point from the perspective of the Greek state (in this case represented primarily by the Hellenic Ministry of Migration Policy) represents direct proof of a lack of state capacity, and subsequently, the state resilience against abrupt changes in the form of migration and refugee flows into Europe. The Greek state already weakened in its fiscal and administrative capacity due to economic upheavals throughout the first half of the 2010s, lacked the necessary capacity to implement its policies when governing side effects from the migration-induced social tipping points, namely overcrowding on the Greek island of Lesvos in the Moria reception camp. The combination of the wrongly defined policy goals, improper implementation of the correctly formulated goals with regard to the asylum management in the Moria camp, and the Greek administrative vulnerability (Illiadou, 2019) resulted in human catastrophe. Migration and refugee flow into Europe in the context of the weak resilience of the Greek state resulted in overcrowding of the Moria camp where at one point over 20,000 migrants, refugees, and asylum seekers lived in a place originally constructed for 3000 people. State failure to govern and be resilient against these strains tipped over the local environment one more time and contributed to the emergence of other social tipping points. Thousands of people living for months in overcrowded deplorable conditions and lacking proper shelter, access to hygiene, health care, food, and even water, in combination with some external amplifiers (e.g., COVID-19; Szabó & Jančovič, 2020), resulted in the camp burndown.

6.1.3 Non-state Actors' Resilience in Moria Camp Burnout

According to the conceptual framework we presented in Fig. 4.7, if the state lacks the necessary capacity and resilience to be able to govern and withstand the disturbance from a social tipping point, other non-state actors might step in and help the state out with their respective capacities and resilience to face these disturbances. Let us, therefore, take a look at how transnational (the EU), private (NGOs), and local (municipalities and local communities) non-state actors behaved in the context of this social tipping point induced by migration inflows, what was their interaction with the state capacity, and how their respective resilience performed when facing the impacts of the European migration and refugee crisis following 2015, in particular, the Moria camp.

After the Moria camp was destroyed in a fire on the night of September 8 and 9, 2020, Ylva Johason (EC Home Affairs Commissioner) firmly declared there would be "no more Morias," accepting some of the blame for the failure of the European migration and asylum policies embodied in the Moria camp burndown (Scipioni, 2017; Gordon & Larsen, 2021). The EU's role in governing the migration and

[5] In June 2021, four Afghans were sentenced to 10 years in Greek prison for intentional arson (BBC, 2021).

refugee crisis is undeniable as the authority over the European asylum system incrementally shifted toward EU agencies as the crisis escalated, especially in the context of hotspots such as Moria (EP, 2016). For instance, fingerprinting proved to be critical to the functioning of the Dublin system and the EC did not hesitate to start infringement proceedings against the peripheral countries (e.g., Greece) if the precautions were not properly and promptly implemented (EC, 2016a). This became a policy priority when it came to hotspots. Naturally, the role of the EU in asylum management and support of peripheral countries such as Greece was relatively significant. For instance, the EU created an EU Border and Coast Guard (EBCG) out of Frontex or an EU Agency for Asylum both of which aimed at promoting the asylum management capacity of its external borders. When the situation at its borders in Lesvos drastically deteriorated, the EU even recommended reinstating, first time in its history, internal border controls to address deficiencies in the external border control in Greece (EC, 2016b). Another significant step was that the EU also provided relatively unprecedented levels of funding to peripheral states hit by the migration and refugee wave, unfortunately, with no structural and lasting solution (den Hertog, 2016).[6] On the contrary, there remains to be a great divide between what is needed to secure proper external border protection and functioning asylum management and what the EU can provide given its limited fiscal capacity.

Emmanuel Goué, MSF head of mission in Greece, declared that "the EU and Greek authorities continue to rob vulnerable people of their dignity and health, seemingly to deter others from coming" (MSF, 2019), even though the EU was expected to contribute to the Greek state resilience the most. In fact, following the implementation of the EU–Turkey statement, the Greek state and its various branches (e.g., the Greek police, the Greek army, the Greek Coastguard, and the Greek Asylum Service) worked closely with several EU agencies (e.g., the EU's Law Enforcement Agency, EBCG, or the EU's Judicial Cooperation Unit, the European Asylum Support Office) on turning the Greek island of Lesvos into a securitized and militarized space of governance (Illiadou, 2019). The joint EU and Greek policies trying to govern the newly emerged social tipping point in 2015 have often been criticized for violating human rights and asylum standards, such as designating Turkey as a safe third country where asylum seekers can be returned. According to Karamanidou (2021), the case of Greece subsequently demonstrates how interactions between the European system of migration and asylum governance and the particularities of the domestic context, such as low resilience, can result in asylum laws and policies contradicting migration and asylum justice. So once again, the EU was not completely innocent in (co)creating and maintaining conditions for the Moria camp burndown. Subsequently, despite Johansen's firm declaration of no more Morias, 6 days after Moria burned down, a new EU-funded camp was opened to replace the burnt-down, Kara Tepe, where thousands of migrants and refugees from Moria were allocated. Moria 2.0 tends to be widely regarded as even worse

[6] It was reported that the sudden rise in refugee inflows meant a nearly twofold increase in the entire EU allocation to the Asylum, Migration, and Integration Fund for the period 2014–2020 (Scipioni, 2017).

than Moria and in spite of the EU's declaration not to ever let something like the burndown of Moria happen again, it openly constructed the same replica of Moria a couple of miles away, risking the same fate (Digidiki & Bhabha, 2020). Although crucial stakeholders had hoped that the disaster in Moria would represent a wake-up call and trigger new negotiations at the EU level for a permanent relocation mechanism, the new Pact on Migration and Asylum did not address almost any of the most pertinent issues. On the contrary, given the pact's emphasis on the externalization of migration control and humanitarian responsibility to third countries, the pressure on the EU's periphery such as Greece is likely to increase. Digidiki and Bhabha (2020) observe that the new pact will further fortify Europe and turn its peripheral countries into de facto prisons for migrants, asylum seekers, and refugees, sowing the seeds for another social tipping point.

One of the main arguments of Gordon and Larsen (2021) is that the purpose of the deplorable and inhuman condition in Moria was to deter migrants and refugees from ever coming to Greece – "a purposeful policy of neglect intended to act as a deterrent within a wider system of exclusionary border practices." Although we still think that the primary driver of the Moria catastrophe was insufficient resilience on behalf of the state, there might still be some truth to it. As the Greek state realized that it lacked the capacity and resilience to govern and withstand the migration-induced crisis, it attempted to avert future migration waves by trying to deter the migrants and refugees by artificially co-creating unbearable conditions. For instance, some participants in the camp observed a purposeful lack of cooperation between state organizations (responsible ministries) intending to cement poor conditions in the camp. Not only that, the Greek police and authorities managed to create a very hostile environment for NGO workers, making sure the job of NGOs became more difficult as they wished to force them out (Gordon & Larsen, 2021). Still, many humanitarian workers and NGOs persevered and became even more motivated to continue working in the camp, trying to fill the gap resulting from a lack of state resilience and the inaction of the state with their respective capacities, especially as the situation escalated dramatically in 2015. For a long time, humanitarian services in Moria were mainly provided by NGOs and informal networks of volunteers and solidarity organizations (Karamanidou, 2021). Here, we can observe that the respective capacities of private non-state actors, in this case, the NGOs, complemented the lack of state capacity to govern the strains of migrants, refugees, and asylum seekers in Lesvos and were doing so, unless the state allowed angry mobs to force some of the NGOs out.

As the Moria camp embodied a symbol of inhuman conditions, disrespect for human dignity, and failure of the Greek state and to an extent the EU, there were numerous clashes between the locals and local authorities and the Moria camp, its residents, and humanitarians. There have been recorded cases of the obstructions of import of accommodation by the Lesvos mayor, who prevented freighters from unloading huts for Moria hotspots, or the refusals to allow building of a kindergarten (Gordon & Larsen, 2021). Few openly violent clashes occurred between local permanent residents of Moria and migrants or humanitarians. These clashes had been accompanied by violent protests and vigilante attacks by various far-right groups. Eventually, the Greek island of Lesvos became a hub of right-wing extremism due to

Moria (TRT, 2020). For instance, following the decision by Turkey to open its borders to Greece on March 29, 2020, allowing thousands of refugees and migrants to cross the border from Turkey to Greece, far-right extremists started attacking migrants, humanitarians, medical staff, and even journalists (Gordon & Larsen, 2021). As a feedback mechanism, this also put many humanitarian actors off and finally made them leave the island for good, bringing to a halt many of the humanitarian projects (ITV, 2020). As we theorized, this is a good example of how non-state actors can enhance the state's capacity and resilience, however, only up to a point. If the state lacks resilience against social tipping points too much, even additional capacity and resilience from non-state actors do not have to suffice, on the contrary, it can deteriorate even further alongside the state's deterioration of resilience and contribute to the emergence of other social tipping points. This is precisely what happened as the departure of crucial humanitarian personnel due to attacks by the far right groups left the refugees and migrants in Moria, especially the most vulnerable among them, exposed to even harsher conditions as essential provisions and services were taken from them (Gordon & Larsen, 2021). Unhuman living conditions, perceived lack of protection by the local authorities, and chaos prevailing in the camp eventually triggered a vigorous response from the locals and the far-right extremist groups made the people trying to help to avert the humanitarian disaster leave, further worsening the situation in the camp with regard to food, health, and hygiene security.

It was argued that the state is a primary actor responsible for acting when a social tipping point emerges and success in governing these abrupt changes is determined by the state's capacity and resilience. If the state lacks proper capacity and is not endowed with the necessary resilience, non-state actors can step in and try to support the state with their respective capacities and resilience. We have seen that this is precisely what happened in the case of Moria with the NGOs. Since day 1, the humanitarians, volunteers, and organized NGOs were often the only providers of humanitarian services and helped to save the state from being quick and absolute when the migration and refugee crisis hit in 2015. The same applies to the EU, which also tried to support the Greek state with its administrative, legal, and fiscal capacity. On the contrary, we should not forget that the EU is also co-responsible for the creation of unhuman and deplorable conditions in the Greek hotspots and keeps doing so even today, after the Moria camp burndown. Moria is also a very transparent case of how the non-state actors can fill some of the capacity and resilience that the state is lacking, however, only up to a point. When the state resilience is excessively unsatisfactory, other amplifiers, such as general discontent with the hygienic conditions in the camp, diseases, and riots, can trigger a backlash from the local communities. In this case, we could notice how at one point local communities and municipalities, in collaboration with the far-right extremists, started openly opposing and attacking not only the residents of the camp, but the humanitarian staff and the employees of the NGOs as well, eventually forcing them out, further weakening the societal capacity and resilience against social tipping points. The Moria camp case is a radiant example of how a state with very weak capacity and resilience cannot be saved from the occurrence of social tipping points even with the help of other non-state actors.

6.2 Tatra Mountain Bark Beetle Outbreak

6.2.1 Identification of the Climate-Induced Social Tipping Point

Tatra National Park (TANAP) located in northern Slovakia represents one of the oldest transboundary protected areas in the world (declared with the Act No. 11/1949 Coll.), an important ecological hotspot, and at the same time a popular tourist destination in the Central European region. The daily visitor count can go as high as 50,000 people during both the summer and winter seasons. TANAP also serves as a refuge for several key animal species of European diversity, foremost great predators such as wolves, brown bears, and European lynx. The mountain range serves as a water tower with many springs and underground freshwater reservoirs, as well as an important supplier for the wood processing industry.

Topographically, the area stretches over the most elevated part of the Carpathian Arc with the highest point of Gerlach Mountain (2655 m). According to the data of the Institute of Nature Conservation, TANAP's total area equals more than 73,800 hectares and is divided into intervention (approximately 42.8% of total area) and nonintervention (approximately 57.2% of total area) management zones. The protective zone extends over 20,703 hectares. Approximately 75% of the area is covered by forest (55,350 hectares). Mature *Picea abies* forests dominate the subalpine zones (Zielonka & Malcher, 2009) that are mixed with European larch, Arolla pine, and a few broadleaved tree species. The forests have a high degree of naturalness in their upper elevations even though only 3% of TANAP can be considered a primeval forest, 13% a natural forest, and 13% a semi-natural forest. The rest 23%, 30%, and 18% are considered slightly altered, significantly altered, and utterly altered forests, accordingly (Fleischer et al. 2009).

Like other coniferous forests in Europe, TANAP has also experienced severe and intense windthrow disturbances that occurred in 2004. According to Bale et al. (2002) or Netherer and Schopf (2010), natural disturbances such as windthrows are strongly climate-sensitive and therefore likely to be distinctly affected by climate change. Especially, European forests appear to be climate-sensitive (Schelhaas et al., 2010), and even though small-scale disturbances are quite common, large-scale disturbances do not tend to be considered to be a natural part of Central European forest dynamics and result from a decreased resistance of forests due to air pollution and climate change (Jonášová et al., 2010). It is now widely accepted that intensifying disturbances such as windthrows are expected to be among the most detrimental impacts of climate change on the services that forest ecosystems provide to society (Lidner et al., 2010; Seidl et al., 2010).

On November 19, 2004, TANAP experienced a very severe windthrow disturbance (windstorm "Elisabeth"). The storm culminated at the maximum speed of 230 kph at timberline (1480 m above sea level) and left more than 12,600 hectares equaling 2.8 mil m^3 of forest uprooted what has been recorded as the largest uprooted volume in TANAP history ever. Once this threshold was passed and the Elisabeth windstorm triggered the abrupt nonlinear change, it led to the overturn of the SES, in this case, the Tatra Mountain forestry. Crossing the threshold started a self-reinforcing feedback mechanism and resulted in a qualitatively, and irreversibly different state of the SES. According to Konôpka et al. (2021), the windstorm hit the northern and central regions of the country. The epicenter of the forest destruction was, however, in the High Tatra Mountains and the Podtatranska basin, which means that most of the disturbed area was part of the TANAP. Naturally, there are numerous possible consequences of such natural disturbances tipping over the forest ecosystems, something which we refer to as side effects (Fig. 5.1). The most common ecological legacies of the windthrows are dead woods and subsequent bark beetle outbreaks on and around the blowdown stand (Jonášová et al., 2010). The bark beetle outbreaks arise, in general, from uncleared trees. The biggest problem though is that the depletion of windthrown trees with low resilience drives bark beetles to attack even healthy trees and typically, a bark beetle outbreak starts to develop from 1 to 3 years after the windthrow (Havašová et al., 2017). It is, therefore, very crucial for the state and non-state actors to build sufficient capacity and inter alia resilience securing that such side effects from windstorm-like social tipping points are mitigated and under control. It is estimated that in the areas with the highest level of protection under TANAP's forest management rules and regulations, around 165,000 m^3 of damaged wood was left uncleared (Nikolov et al., 2014). Subsequently, the bark beetle-led tree mortality that followed the windthrows happened to have outgrown the storm-led forest loss in the TANAP highest protection areas (Fig. 6.3). For instance, Javorova valley (marked as Tatranska Javorina),

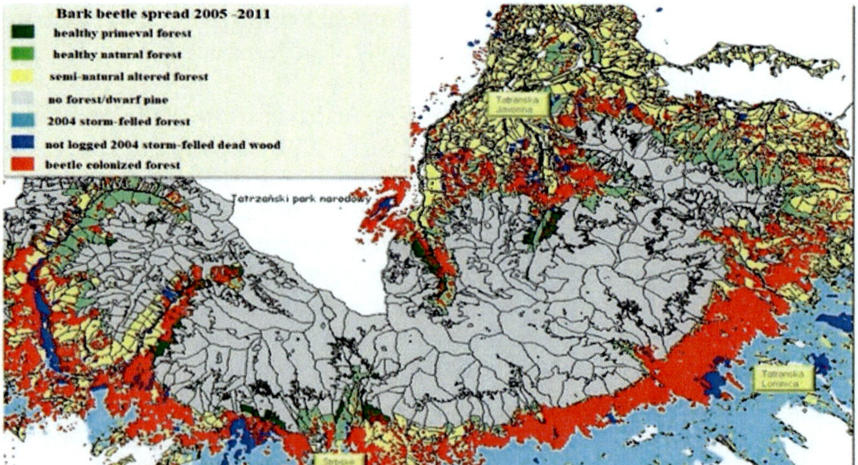

Note: data state forest enterprise of TANAP, degree of naturalness.
Source: Fleischer et al. (2009).

Fig. 6.3 Bark beetle infestation in Tatra National Park, 2005–2011. (*Source*: Fleischer et al. 2009) *Note*: **data state forest enterprise of TANAP, degree of naturalness**

one of the TANAP core protection zones that include old-growth primeval up-to-300-year-old forests, experienced more than 50-fold increase in spruce bark beetle-led mortality (from 3.29 hectares up to 175 hectares) in 2002–2015.

Furthermore, the bark-beetle eruption in 2005–2015 additionally contributed to many other major disturbances to ecosystem services, particularly to water catchment and soil stability (Strzyżowski et al., 2018), soil nutrient (Šimonovičová et al., 2019), carbon catchment (Fleischer et al., 2020), and organic carbon stocks (Don et al. 2012). According to Strzyżowski et al. (2018), the sediment flux rate for the whole area of TANAP in 2013 reached 3.55×10^{-4} $m^3/m^{-1}/year^{-1}$, while the mean sediment flux based on the 48-year data was at 2.76×10^{-5} $m^3/m^{-1}/year^{-1}$ level. It was estimated that three fourths of the TANAP area is at high risk of erosion with 5–15 mm yearly soil loss. Moreover, the data of State Forests of TANAP show that while in 2008, Tatra Javorina recorded 256.7 mm of rainfall in 4 days accompanied by one small landslide in the glacier moraine, in 2014, the rainfall was significantly lower – 165.6 mm, and the number of landslides in glacier moraine significantly erupted to 19 in 2014.

The loss of forest cover in the area was also linked to the loss of air quality (Fleischer et al., 2020). The climatic spa areas in TANAP focusing on healing respiratory syndromes such as Vysne Hagy and Strbske Pleso reported higher carbon volumes. The same authors argue that 1 hectare of full-grown trees stores up to 120 tons of carbon. Last but not least, the number of TANAP visitors from 2005 through 2011 declined which some evidence (Arnberger et al., 2018) linked to the loss of visual satisfaction. Yet, less evidence has been put forward for other economic losses of TANAP 2004 windthrow such as the assessment of lost forest asset value or quantity and quality of marketable timber products (Morris et al., 2018).

6.2.2 *State Resilience to the Tatra Mountain Bark-Beetle Outbreak*

So far, we have identified a case of climate-induced STP, the calamitic windthrow affecting forest resilience in Slovak Tatra Mountain that resulted in various side effects, namely the bark-beetle pest outbreak the state had to deal with. According to Makrickiene et al. (2019), forest ownership in Slovakia is 40% state, 10% private, 20% municipal, 11% agricultural cooperatives and church, and 19% either unknown or under restitution. Of the forest zones with prescribed forest management regimes, strict natural reserves represent merely 2% of the area, special purpose with restricted management only 10%, 17% protective, and 71% commercial use. TANAP, a nonintervention zone of forest management in Slovakia and the area that was the most impacted by the Elisabeth windstorm, is under the direct administration of the state-funded Správa Tatranského národného parku and therefore directly governed by the state. State with its capacity and resilience shall, therefore, be at the forefront of forest management in TANAP in Slovakia. As we already outlined before, facing the unexpected Elisabeth windstorm in November 2004, the state decided to leave 165,000 m^3 of damaged wood uncleared, contributing to the occurrence of the bark beetle outbreak (Nikolov et al., 2014).

To determine the state's resilience to withstand disturbances stemming from STP, we first decided to quantify the overall loss in the TANAP's forestry due to the pest epidemic. In compliance with Kovalčík et al. (2018), who suggest calculating the value of forest lost as a reduction of the basic value of forest stands, we estimate the loss in forest stands. This approach adheres to our view of resilience that takes into account that society is an integral part of the ecosystem, and social–ecological system, and considers the losses to the economy as an inherent part of ecosystem disturbances. The adverse effects of forest loss on aggregate product and redistribution of wealth are therefore subject to the scale of the disturbance.

The economic value of forest stands and forest lots (deforested land) is determined by the Ministry of Justice of the Slovak Republic in its Regulation No. 492/2004 Coll. In line with this approach, we calculated the loss in forest stands value (Table 6.1). The relative value of forest stands to forest lots is directly proportional to the level of protection. Thus, the loss of forests in higher protection zones 4 and 5 incurs relatively higher costs than those in low protection zones. In particular, the forest stands in the lowest protection zone are only 6.73 times more expensive than forest lots, while the value of forests in protection zone 5 is 12.62 times higher than the value of deforested land at that level.

Given the volume of bark beetle-led tree mortality rate according to Nikolov et al. (2014), we estimate that TANAP has lost 7544.3 million euros in terms of forest economic value. Kovalčík et al. (2018) also suggest adjusting the losses to forest stand value by defoliation. The economic value of defoliated stands is yet again determined by the Regulation of the Slovak Ministry of Justice No. 492/2004 Coll. Thus, the adjusted total losses to the TANAP economy represent 7083 million euros, which is close to the estimates by the Ministry of Agriculture and Rural Development (MARD, 2011). It is fair to say that over 7 billion euros and 2.5 million m^3 of spruce destroyed between 2005 and 2010 over 120 m^2 due to the bark beetle outbreak are unparalleled numbers (Økland et al., 2016).

Based on the information presented so far, we argue that the damage in forest value culminating in over 7 billion euros occurred mainly due to mismanagement of STP on behalf of the state, indicating weak resilience of the state. It has long been recommended that windthrow monitoring and management in spruce-dominated stands is crucial for successful forest management and that damaged wood due to

Table 6.1 Comparing average prices of forest lots and forest stands according to the level of protection (€/ha)

Level of protection (LoP)	Average value – forest lots		Average value – forest stands		Relative average value – forest stands/forest lots	
LoP1	824.55	113.6%	5546.74	110.0%	6.73	96.8%
LoP2	713.20	98.2%	4950.28	98.1%	6.94	99.9%
LoP3	446.46	61.5%	3509.80	69.6%	7.86	113.1%
LoP4	419.54	57.8%	4216.67	83.6%	10.05	144.6%
LoP5	218.88	30.1%	2761.63	54.8%	12.62	181.6%
Total	726.00	100%	5043.69	100%	6.95	100%

Source: Authors' calculations based on data from the National Forest Centre the windstorm should not be left uncleared because of the possibility of the bark-beetle outbreak (e.g., Angst et al., 2012). Nikolov et al. (2014) also implicitly argue that the extensive damage resulting from the bark-beetle epidemic was a result of the lack of state capacity and resilience, even though they refer to it as a lack of monitoring and forest management. The authors argue that the state failed in three steps. First, since the peaks of the bark beetle population densities in mountain (spruce) forests culminate in the third summer after a windstorm, all damaged wood should be removed in the first 2 years following the windthrow, whereas wind-felled trees in the vicinity of infestation spots are recommended to be cleared first. This observation is also supported by Økland et al. (2016). As was already pinpointed numerous times, this has not happened. Second, a 300 m phytosanitary buffer zone from the epicenter should have been secured. Third, even if there were places where it was impossible to remove the damaged wood due to its inaccessibility in the mountain or due to the fact that certain zones were within the protected area (e.g., Crofts et al., 2005), a buffer zone and adjacent managed sites should have been built. Neither of these policy precautions stemming from the empirical evidence has been done. Quite the contrary, the state opted for minimal salvaging of wind-felled trees culminating in the bark beetle outbreak, where total bark beetle damage during the first 5 years turned out to be three times the damage than the volume of uncleared wind-felled trees (Nikolov et al., 2014; Økland et al., 2016).

As we can see, the exclusion of post-disaster management had adverse consequences for forest stability as it turns out clearing windthrows is the most effective control measure. Even relatively small areas of uncleared windthrow trees and initial bark beetle infestation spots can trigger extensive bark beetle outbreaks with immense costs due to weak state capacity and resilience. If we delve into the reasons why the state failed to avert even the most imminent losses, we can identify two reasons. First, the state lacked the capacity in an administrative and infrastructural domain to mitigate the impacts resulting from the climate-induced STP which in turn incapacitated the state in its resilience vis-à-vis the side effects stemming from it. In a nutshell, the state was unable to send out experts and administrative staff (low efficiency of civil service) that would be able to deal with the issue adequately. At the same time, it also lacked resources to do so (forest machinery and equipment and means of transportation), both fiscally and militarily. Second, even if the state deliberately opted for the noninterventionist policy concerning the wind-fallen wood, it still failed to select the most suitable policy as a result of low state capacity in administration. Especially when it has long been known that wind disturbances and subsequent windfallen wood might end up triggering bark beetle outbreaks (Hlásny et al., 2021). Mezei et al. (2014), analyzing bark beetle outbreak in Slovakia before 2000, concluded that "precise bark beetle control can significantly slow down the speed of stand break-up," so the Slovak authorities should have already had some benchmark to which they could turn to. Administrative staff with higher human capital would be undoubtedly able to identify the harmfulness of the laissez-faire forest management approach, especially when it is known that this kind of approach can have adverse impacts beyond forest resilience, for instance, on local communities, economy, and society.

If we assume that the state resilience to absorb adverse effects stemming from the windthrow is a function of the disturbance size and local economic conditions, our unit of analysis is still SES after all, it is plausible that the disturbance in the context of the dependence of the local economy on timber has also affected the environment surrounding TANAP, namely wood processing industry. It is assumed that when the wood processing industry faces natural disturbance in the form of the bark beetle outbreak, the price of timber drops and the sector faces medium- to long-term timber supply outages. Therefore, in accordance with Kovalčík et al. (2018), we estimate the loss in forest stands and subsequently the cost of three mortality-led outages in timber supply. The immediate and 5- and 10-year-later socioeconomic costs of tree mortality-led outages in timber supply can be found in Tables 6.3 and 6.2, respectively. The immediate impacts on the local economy appear to be quite sizeable given the timber volume left uncleared, which further contradicts the laissez-faire forest management approach (Table 6.3). Hlásny et al. (2019: 35) confirm that "the outbreak has had catastrophic impacts on the regional forestry economy," as the salvaged timber could not be sold but was accumulated in

Table 6.3 Costs of tree mortality-led outages in timber supply, 2010 and 2015 TANAP Slovakia, 2017 prices

Indicator	Unit	2010	2015
Revenues	mil. €	48.47	90.55
Gross value added		17.64	32.96
Net value added		13.58	25.38
Profit		2.08	3.89
Tax revenues		4.27	7.99
Health and social security revenues		3.37	6.30
Personal income		8.70	16.25
Employment	persons	852	1591
Workforce		604	1128

Source: Authors' calculations based on data from the National Forest Centre

Table 6.2 Costs of uncleared windthrown trees in protection zones, cumulatively 2004–2006 TANAP Slovakia, 2017 prices

Indicator	Unit	FPA 3–4	FPA5	Total
Revenues	mil. €	232.95	90.88	323.83
Gross value added		84.79	33.08	117.87
Net value added		65.28	25.47	90.75
Profit		10.02	3.91	13.92
Tax revenues		20.54	8.01	28.56
Health and social security revenues		16.20	6.32	22.52
Personal income		41.80	16.31	58.11
Employment	Persons	2902	1132	5690
Workforce		4093	1597	4034

Source: Authors' calculations based on data from the National Forest Centre storage yards and acted as an additional breeding source for bark beetles. Their data support our findings that as roundwood was sold as fuelwood causing income losses for private forest owners, and an indirect decline in demand for forest workers, the regional forestry economy deteriorated.

Comparing the data with the 5- and 10-year checks, we see that the losses doubled (Table 6.3). Similar estimates were also reported by Kovalčík (2018) and the Ministry of Agriculture and Rural Development of the Slovak Republic. These data show that even climate-induced STP such as windthrows and subsequent bark-beetle outbreaks due to mismanagement on behalf of the state can locally result in an immense cost that could have been avoided if only the state had been more resilient and had greater state capacity.

6.2.3 Non-state Actors' Resilience to the Tatra Mountain Bark Beetle Outbreak

Forest management in Slovakia mostly relies on state and state-funded companies, namely the state authorities (e.g., Ministry of Agriculture and Rural Development), state-funded organizations (e.g., National Forest Centre in Zvolen), and state-owned companies (e.g., Forests of the Slovak Republic). In general, the management of forest disturbances in central European countries is strongly influenced by traditional forest management approaches with insufficient focus on adaptation to climate change, including the emergence of pests. As we can see, the traditional methods of pest management in combination with the state's lack of capacity and resilience to deal with climate-induced beetle outbreaks can generate great environmental, economic, and societal losses. According to our findings, we have to agree with Hlásny et al. (2019) who argue that national crisis plans for cross-sectoral cooperation in economy, transportation, forestry, or public safety are still insufficient for early and effective mitigation of large-scale disturbances. We have established that a lack of state resilience in forest management allowed malign impacts resulting from the bark beetle outbreak on its society. Let us look at the non-state actors and their involvement in this social tipping point.

Although windstorms leading to pest outbreaks do not generally acknowledge state boundaries, transnational resilience in forest management in Europe is absent. Since the EU treaties make no specific reference to forests, there is no common forestry policy, and it thus remains primarily a national matter. On the contrary, there are still attempts to unify forest management and resilience across the member states, although Aggestam and Pülzl (2018) argue that without coordinating collective EU goals and gathering strong political support, it is impossible to achieve coherence for EU forest-related policies. As a result of this, there was no immediate and tangible support from the European institutions. Still, in the aftermath of the windstorm in 2005, the International Union for Conservation of Nature visited TANAP and pointed out possible sources of financial support from the EU, namely the EU Rural Development Regulation for positive support given the status of TANAP as a Natura 2000 site (Crofts et al., 2005). Eventually, Slovakia received over 5 million euros from the EU as an act of solidarity although, in June 2007, the EC officially started an infringement procedure against the Slovak Republic due to alleged violation of the EU directives concerning the Nature 2000 implementation in High Tatra region (MARD, 2011).

Immediately after the windstorm though, numerous local activists, volunteers, and representatives of cities and municipalities started helping to get rid of the waste from the windstorm, build up destroyed infrastructure, and resolve all the immediate problems stemming from the calamity. Direct elimination of all the damages lasted a couple of months (ŠOPSR, 2018), and the local non-state sector supported the state in dealing with the climate-induced STP. Since the affected region borders also with Poland, volunteers from the neighboring countries came to help out to mitigate the immediate impact of the windstorm as well.

Environmental NGOs, mostly with antipathy toward the interventionalist approach regarding the consequences of the windstorm, took a relatively aggressive stand against already minimalistic attempts to get rid of the windfallen wood to avoid a potential bark beetle outbreak (e.g., the Central and East European Working Group for the Enhancement of Biodiversity, Greenpeace, and forest protecting association Wolf), and launched an ongoing legal battle against the Slovak state attempting to remove the windfallen trees from the TANAP (MARD, 2011). We are not going to debate the merits of the noninterventionalist/interventionalist approach toward the natural disturbances causing bark beetle outbreaks, we have already done that before. We just want to point out that the neighboring Czech Republic, which recently faced large-scale windthrows and also subsequent bark beetle outbreaks, managed to mobilize key state and non-state stakeholders and led a constructive dialogue between government and environmental NGOs despite their contradicting views. As a result, the overall economic, environmental, and societal losses in the Czech forests appear to be fewer than in Slovakia (Hlásny et al., 2021). State and non-state resilience, therefore, requires a synergy from both sides.

The case of the Tatra Mountain bark beetle outbreak reveals, similar to the Moria case, that a lack of state capacity incapacitated the state when it was supposed to face the direct and indirect effects of the climate change–induced STP. A lack of administrative, legal, infrastructural, fiscal, and military capacity meant that the state was unable to dispatch experts and administrative staff, law-enforcement resilient units, or forest machinery and equipment to face and withstand the disturbances from the windstorm. In the subsequent step, the state lacked the capacity to opt for the most suitable policies toward forest management and indirectly contributed to triggering the bark beetle outbreak with much more dire consequences in terms of economic costs and environmental damage than the first STP in the form of a windstorm. It has to be pointed out here that even though the non-state actors, namely local municipalities and volunteers, provided their respective capacities and helped the state to be resilient against the windstorm, eventually, insufficient capacity and resilience of the state toward the climate change–induced disturbances caused that even additional non-state actors' resilience was not enough. Moreover, the missing cooperation between the state and non-state actors (mainly NGOs and nonprofit environmentalists) caused the overall resilience of Slovakia, a peripheral EU country, to weaken and contributed to the emergence of the bark beetle outbreak.

6.3 War in Ukraine

6.3.1 Identification of the Geopolitics-Induced Social Tipping Point

The last example of STP in this monograph is dedicated to the geopolitical tensions on the Russo-Ukrainian board culminating into an unprecedented, from the point of view of the twenty-first century, military campaign of one nuclear superpower toward its smaller neighbor. Putting the 2004 Orange Revolution aside, this geopolitics-induced STP started gaining momentum at the turn of 2013 and 2014 following the people's overthrow of President Viktor Yanukovych in Ukraine. On November 21, 2013, President Viktor Yanukovych, relatively unexpectedly, defied the country's previous stance and declared against signing the EU–Ukraine Association Agreement. Yanukovych instead started promoting closer relations with the Russian Federation which ended up infuriating a lot of Ukrainians, especially in the Western parts of the country. By late November 2013, massive protests against the government burst out. As the situation kept deteriorating and the pro-government law-enforcement units attempted to crush the protestors, street clashes erupted in the capital and other cities as well. Even though President Yanukovych ultimately agreed to form a unity government with opposition leaders and called for snap elections on February 21, 2014, it was too late. The next day, the law-enforcement units lost control of central Kyiv, and Yanukovych was forced to flee the city. On the same day, the parliament voted to remove him from office. On February 27, 2014, the opposition forces formed an interim government. At the same time as the anti-Yanukovych opposition officially seized power and unsuccessfully attempted to legislatively repeal Russian its status as an official language for public administration, unmarked Russian military men invaded the Crimean Peninsula. "Little green men," as the masked Russian soldiers were initially labeled by the Russian media, swiftly took over the television station, government buildings, and other strategic infrastructure, such as the Simferopol airport, with no use of violence. This unexpected turn of events happened as the whole world was still focusing on the situation in Kyiv's Maidan. On March 6, 2014, the Crimean lawmakers voted to secede from Ukraine and called for officially joining the Russian Federation. This decision was later confirmed in an unmonitored referendum held on March 16, 2014, with 95% of Crimean participants supporting the decision to become part of Russia. The Russian annexation of Crimea was subsequently finalized on March 18, 2014, as Russia formally incorporated the peninsula into its political system (Karagiannis, 2014). Violating basic principles of international law, the annexation of Crimea was never recognized by the international community.

Crimea, unfortunately, did not represent the only rupture between Ukraine and the Russian Federation. In the aftermath of the 2014 Euromaidan protests and Ukrainian revolution, an armed conflict between anti-government groups of pro-Russian separatists and the new pro-EU government emerged in the eastern part of Ukraine, namely in the Donetsk and Luhansk oblasts, regions that tend to be collectively called the Donbas region. Overthrown President Yanukovych and his Party of Regions had the Donetsk and Luhansk provinces as their strongholds for many years. Naturally, the mostly Russian-speaking regions did not find the pro-EU Maidan movement very positive, and their concerns quickly hardened into widespread militant opposition. Against the backdrops happening in the capital, the demonstrators in Donetsk started voicing their demands to institute a referendum similar to the one held in Crimea and to declare the newly formed Ukrainian government illegal. When the lawmakers refused to accede to the demonstrators' demands and the demonstrators were quickly expelled by special Ukrainian forces, anti-Maidan protesters overran regional administration buildings, city councils, prosecutors' offices, and broadcaster centers in 32 cities. Subsequent development led to the proclamation of the founding of the Donetsk People's Republic (DPR; April 7, 2014) and the Luhansk People's Republic (LPR; April 27, 2014), respectively, and the advent of the firefights between the Ukrainian government forces and Russia-backed rebels (Clarke, 2016).

In spite of peace-making attempts in the form of two Minsk agreements intermediated under the auspices of the so-called Normandy Four – Russia, Ukraine, Germany, and France – peace and stability to the war-torn Donbas region were never fully brought following the start of the conflict in 2014. The first Minsk Agreement was signed in September 2014; however, none of the agreement's 13 articles has been fully implemented. In January 2015, the Russia-backed forces of the self-proclaimed DPR and LNR embarked on an offensive trying to retake territory lost to Ukrainian government forces in mid-2014, which culminated in the second Minsk Agreement in February 2015. Although some progress has been made regarding the exchange of prisoners or partial withdrawal of heavy weapons from the line of contact, the ceasefire has been broken countless times between 2015 and 2020 (Åtland, 2020). In 2019, the presidential election in Ukraine resulted in Volodymyr Zelensky winning the presidency in a landslide victory with a promise to end the conflict in eastern Ukraine and pro-western integrational outlooks (Rohozinska & Shpak, 2019). Not surprisingly, the "silent" Russia-backed military support to secessionist DPR and LPR movements fueled greater enthusiasm for joining not only the EU but also the NATO, and in January 2021, President Zelensky appealed to the US President Joe Biden to let Ukraine join NATO. Russia subsequently began mobilizing its troops near Ukraine's border in the following months of spring 2021, allegedly with the purpose of a training exercise. By the end of 2021, as satellite images revealed that the Russian forces near the border with Ukraine already surpassed 100,000 troops deployed, Russia presented its security concerns calling for Ukraine to be barred from ever joining NATO and thereby posing a security threat to Russia. Since NATO explicitly backed its "open-door" policy regarding the potential accession of Ukraine, the situation escalated until the

Russian troops crossed the Ukraine–Russia border on the night of February 23–24, 2022, openly launching a military offensive – "special military operation" in Vladimir Putin's parlance – toward neighboring Ukraine (Person & McFaul, 2022). The delusional Putin's goal of denazification in Ukraine – antidemocratic regime change – represents a result of the systemic change following the geopolitics-induced STP. The crossing of the Russian military personnel through the Russia–Ukraine border on the night of February 23–24, 2022, or rather Vladimir Putin's decisions that launched this bloody cascade, embodies a threshold. Crossing this threshold generated a nonlinear change that tipped the previously frozen conflict state into a new state, the open war between two European nation-states. Similar to the previous two cases analyzed in this chapter (climate-induced and migration-induced STP), this escalation resulting from the aforementioned geopolitical circumstances also triggered a self-reinforcing feedback mechanism, causing numerous side effects. The side effects are numerous, ranging from a global economic downturn stemming from the disruption of global supply chains (Liadze et al., 2022), a slowdown of global trade (Orhan, 2022), shocks to the stock markets (Ahmen et al., 2022), endangering of the public health (David et al., 2022), or shifting investors preferences (Singh et al., 2022). According to the early estimate, the economic costs of the Russia–Ukraine conflict can amount to 1 trillion euros in 2023, which is about 1% of global GDP. Moreover, the conflict can, directly and indirectly, add up to 3% to global inflation in 2022 and about 2 percentage points in 2023 (Liadze et al., 2022). Among the side effects, Europe has to govern and be resilient against one stand out in particular – the exodus of Ukrainians fleeing the war-torn country attacked by Russia.

As was already outlined in the introduction, emigration of Ukrainian citizens did not begin on February 24, 2022, however, gradually increased over the last three decades following 1991 when Ukraine gained its independence after the dissolution of the Soviet Union. The continuous outflow of people from Ukraine was powered by bad economic outlooks, low trust in Ukrainian authorities, as well as state capture by the economic elites and post-communist oligarchy (Mol et al., 2017). According to Vollmer (2016), around 6 million Ukrainians were living abroad, even before the annexation of Crimea by the Russian Federation and the Kremlin-backed uprising in the Donbas region in 2014. The new political circumstances, naturally, altered the push factors influencing the migratory flows out of Ukraine from economic to mostly security-related. In the ensuing conflict, 1.7 Ukrainians were internally displaced, and Ukraine became third in the asylum applications to the EU. Once the first troops crossed the Russian–Ukrainian border, the exodus of Ukrainians seeking temporary protection from the neighboring EU countries started en masse and most economic and safety-related migration morphed into massive refugee outflows leaving Ukraine due to the international armed conflict. UNHCR reports that as of this writing, over 7.8 million refugees from Ukraine are recorded across Europe, and over 6 million Ukrainian people remain internally displaced due to the war. Table 6.4 displays UNHCR data regarding the number of refugees fleeing Ukraine since the beginning of the Russia–Ukraine war. Besides the Russian Federation, we can see that among the Ukrainian neighbors, the most impacted country remains to be Poland hosting over 1.5 million Ukrainian refugees, almost 500,000 more than the biggest European economy, Germany. Among other European countries mostly hit by the refugee crisis following the Russian–Ukrainian war are the Czech Republic, Western European countries (Italy, Spain, the United Kingdom, and France), and Slovakia. In the subsequent parts of this chapter, we will assess how Poland in the EU's eastern periphery faced the strain of refugees fleeing the Russian aggression in Ukraine.

6.3.2 State Resilience in Managing Ukrainian Refugee Inflows

Poland was one of the most heavily affected countries by the war in Ukraine, mainly due to its proximity to war-torn Ukraine. Poland hosts the most Ukrainian refugees among the European countries (Table 6.4). The influx of refugees from Ukraine caused Poland to exceed 40 million inhabitants for the first time in its history. With

Table 6.4 United Nations High Commissioner for Refugees (UNHCR) data on Ukrainian Refugees following the war in Ukraine

	Refugees from Ukraine registered for temporary protection or similar national protection schemes	Refugees from Ukraine recorded in country	Border crossings from Ukraine
Countries featured in the Refugee Response Plan			
Bulgaria	148,451	51,140	–
Czech Republic	471,481	472,473	–
Hungary	32,850	32,850	1,929,514
Poland	1,544,074	1,544,074	8,349,746
Moldova	–	99,524	726,676
Romania	98,162	103,167	1,695,870
Slovakia	103,941	104,140	1,024,101
Other countries neighboring Ukraine			
Belarus	–	17,787	16,705
Russia	–	2,852,395	2,852,395
Other selected European countries			
Germany	1,021,667	1,021,667	–
Italy	166,467	173,231	–
Spain	158,789	158,789	–
United Kingdom	150,600	150,600	–
France	118,994	118,994	–

Note: – means the data were not available. The Regional Refugee Response plan brings together various stakeholders (UN, NGOs, and other relevant partners) and focuses on supporting host countries in order to ensure safe access to territory for refugees and third-country nationals fleeing from Ukraine

Source: Data retrieved from UNHCR operational data portal – Ukraine refugee situation (Accessed on December 20, 2022)

millions of refugees fleeing their homes in Ukraine and crossing the border into Poland, Poland became the second-largest refugee destination in the world after Turkey and witnessed easily one of the largest movements of refugees in modern times. The scale of the refugee inflows into Poland is such that the population of cities such as Rzeszów, the largest city in south-eastern Poland, has increased by more than 50%, in Gdańsk, the population has grown by almost 35%, and in Warsaw by 15% (Fig. 6.4). Nowadays, every third resident of Rzeszów is of Ukrainian origin (35%), whereas, in Gdańsk, it is 25%, and in Warsaw, it is 13% (Wojdat & Cywiński, 2022).

The strain on Poland was immense from day 1. For instance, already on February 24, 2022, more than 30,800 people crossed the border from Ukraine to Poland, trying to flee the Russian forces attacking Eastern Ukraine and Kyiv. The number of incoming refugees kept rising until it reached levels of 140,000 people on March 6 and 7, 2022 (Wojdat & Cywiński, 2022). Given its limited capacity as a peripheral EU country, the Polish state acted relatively quickly and managed to provide necessary help to the Ukrainians seeking refuge in Poland, in terms of clothing, accommodation, as well as administrative and legal help. Making use of its administrative and legal capacity, Poland mobilized other peripheral EU countries and enforced the equalization of the legal status of Ukrainian refugees with Polish citizens at the EU level (Council Implementing Decision 2022/382 of March 4, 2022). The Act on assistance to citizens of Ukraine in connection with the armed conflict on the territory of that country was passed already on March 12, 2022. The bill was signed on March 26, 2022, by President Duda coming into force retroactively on February 24, 2022. This bill allows the integration of the Ukrainian refugees into the existing health, social assistance, education, and labor law system under the same conditions as it does for Polish citizens (see also part regarding the resilience of the EU as a

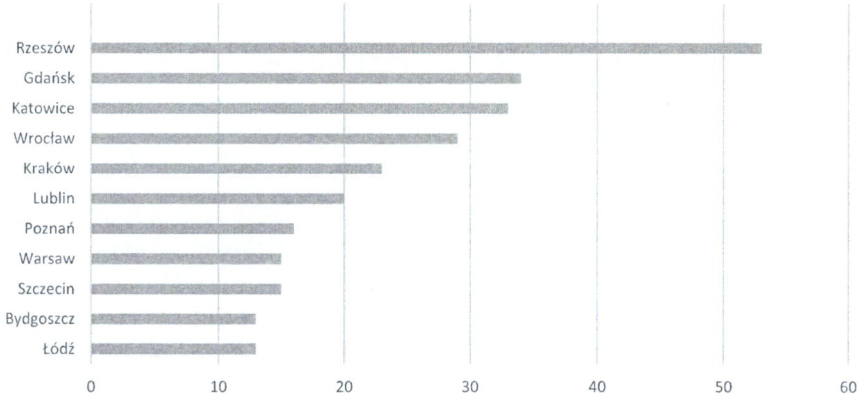

Note: as of 1 April 2022.
Source: Wojdat & Cywiński (2022).

Fig. 6.4 Increase in the population of Polish cities after February 24, 2022. (*Source*: Wojdat & Cywiński, 2022)
Note: **As of April 1, 2022**

non-state actor). For instance, one of the most important rights of a Ukrainian refugee in Poland is the possibility to get a Polish personal identification number (the PESEL) giving theme same access to social, legal, and administrative services the Polish people have. Relatively prompt approval of this bill reveals that the resilience of the Polish state was sufficient to make sure that the integration of Ukrainian refugees into the legal system of Poland is secured. Comparing this case with the Moria camp, it is apparent that this approach represents a completely different experience compared to the treatment of Syrians and Libyans, who after fleeing the war remained locked up in refugee camps (e.g., Moria camp) for years with an almost nonexistent possibility of movement. The combination of low resilience, bad governance, and inhuman conditions in Moria resulted in a humanitarian disaster. Even if they eventually managed to enter Europe, these refugees and asylum seekers did not end up having the same rights as the Ukrainians. The Polish state proved itself to be much more resilient and capable when facing the refugee flows from Ukraine.

Besides cooperating with the EU, the Polish set up a proper administrative and legal environment for the NGOs and municipalities. For instance, based on the provision of Art. 12 para. 4 of the Act on Aid for Ukrainian citizens, any local administrative unit may have on their initiative and within the scope of their available funds provided aid to Ukrainian refugees, which encouraged the help provided by local authorities (Ociepa-Kicińska & Gorzałczyńska-Koczkodaj, 2022). Furthermore, the state also established the so-called Aid Fund, a fund providing resources for all activities and projects related to integrating Ukrainian refugees. The fund is operated by a Polish development bank, Bank Gospodarstwa Krajowego, whereas resources from the fund are also provided to the self-governments in Poland to help them deal with the refugee strain. Help to municipalities was also intermediated by another important policy – concessions that waived the constraints stipulated in the Act on Public Finance (Art. 128 para. 2) and the Act on Revenues of Local Administrative Units (Art. 42 para. 3). The amendment also waived the upper limit on subsidies that can be provided to local administrative units to compensate for the day-to-day outlays and long-term investment to schools and educational institutions helping to integrate Ukrainian students. Also, aid expenses have been excluded from the computation of the ratio of day-to-day revenues as well as the constraints regarding the debt repayment amount (Art. 242–243). Various other important legal steps have been taken to allow the local authorities to provide the necessary help for Ukrainian refugees more efficiently, which concern, for example, amendments in the Public Procurement Law and the Tax Code (Ociepa-Kicińska & Gorzałczyńska-Koczkodaj, 2022). In the case of the amendments to the Public Procurement Law, the state lawmakers allowed the local authorities to outsource the implementation of public tasks to NGOs and other civil society organizations (e.g., trade unions) without the need to announce an open tendering procedure. Public health can be also outsourced without the need to hold a bidding contest. When it comes to the amendments in the Tax Code, numerous tax reliefs were introduced to those helping Ukrainians, covering both personal income tax and corporate income tax and VAT. Moreover, a zero VAT rate was introduced for goods and services that constitute aid for Ukrainian refugees. Additionally, several tax deduction possibilities for donations (e.g., financial donations, free-of-charge services, health care, and blood donation) for Ukraine were allowed (Ociepa-Kicińska & Gorzałczyńska-Koczkodaj, 2022). All these policies fit very well into the state's intentions to encourage private and local actors to use some of their capacities and resilience to help the state deal with the wave of Ukrainian refugees and promote public–private as well as public–local partnerships.

The Polish state was also relatively prompt in creating reception points for Ukrainians (36 reception points as of March 31, 2022). Contrary to the processes observed in Greece, these places served as a place to provide the Ukrainian refugees with necessary information regarding their right of residence or application for international aid (in Ukrainian). Refugees were offered a rest, meal as well as medical and psychological help and were allowed to leave any time they wanted as it was not required of Ukrainians to register at these points to enter the territory of Poland (Gov.pl, 2022). Among other things, this was also a very important aspect of why this strain did not overwhelm the Polish state for the time being.

The overall resilience of the state in Poland and subsequent successful governance of the consequences stemming from the geopolitics-induced social tipping point in the form of refugee flows can be primarily attributed to two factors. The first is the collaboration of the EU and member states with the apparent lead by Poland in introducing a special protection status for Ukrainian nationals and their spouses and equalization of their status with the domestic citizens. This laid the groundwork for a drastically different approach to refugee protection and refugee governance than in the case of Greece. Second, since the Polish state understood the lack of resilience in several aspects necessary for successful governance of the social tipping point, it acceded to helping the non-state actors relatively quickly, providing fiscal and administrative help to the NGOs and municipalities. As a result, numerous side effects from the geopolitics-induced social tipping point, such as a massive strain of the Ukrainian refugees on the Polish–Ukrainian border, that managed to penetrate the wall of the state resilience could be handled by the non-state actors and their respective resilience. Let us now take a look at how the non-state actors succeeded in managing these refugee flows.

6.3.3 Non-state Actors' Resilience in Managing Ukrainian Refugee Inflows

Although the Polish state was rather quick in governing the Ukrainian refugee influx, it lacked state capacity in several aspects. Some experts argue that much more could have been done in terms of preparedness and criticize that the state was not very well equipped in terms of pre-crisis early warning systems (Wojdat & Cywiński, 2022). Although it is argued in the previous chapters that STP are ungovernable due to their abrupt and nonlinear nature, the state had at its disposal numerous intelligence reports warning of the possibility of Russian aggression and could have therefore invested in strengthening its resilience. The state's role in governing the side effects stemming from the Russia–Ukraine war, namely the refugee influx, had to be complemented by non-state actors, and in our case, mainly the EU (transnational actor), municipalities, and local self-governments (local actors), but also civil society organizations and NGOs (private actors).

Given its institutional significance, a very important role in governing this geopolitics-induced social tipping point was played by the EU. Some refer to the EU's response to the side effects stemming from the Russian war against Ukraine as unprecedented in scope and unexpected speed, displaying a rare unity among EU member states and marking the EU's geopolitical awakening (Scheffer & Weber, 2022). Following the urgent call of the home affairs ministers on March 2, 2022, the Commission immediately proposed to activate, for the first time since its introduction in 2001, the Temporary Protection Directive 2001/55/EC (Article 5).[7] The Council subsequently unanimously adopted on March 4, 2022, the decision 2022/382 regarding the equalization of the legal status of displaced persons from Ukraine with the EU citizens. The Ukrainian beneficiaries have a right to a residence permit and access to the asylum procedure, employment, suitable accommodation or housing (see also Safe Homes Guidance), social welfare, medical care, education, and banking services based on the adoption of this decision. Not only that, but the displaced Ukrainians also fleeing the war are entitled to move to another EU country before the issuance of a residence permit and to move freely within EU countries other than the member state of residence for 90 days within a 180-day period after a residence permit in the host EU country is issued (EC, 2022c). To coordinate the implementation of the Temporary Protection Directive, the EC (coordinated by the Directorate-General for Migration and Home Affairs) also set up the Solidarity Platform, which allows for monitoring the needs identified in EU countries and coordinating the operational response. The platform, among other things, provides a general forum for discussion to support the implementation of the EU's 10-Point plan, which turned out to be crucial for the coordination of EU member states' efforts to help Ukrainians fleeing the war. The main advantage of the platform is that it brings together EU countries, Schengen Associated States, various EU-affiliated agencies (e.g., EU Agency for Asylum, Frontex, and Europol), IOM, UNHCR, and Ukrainian authorities (EC, 2022c, 2022d).

Other EC programs also played a crucial role in assisting EU member states with border management, asylum registrations, information-sharing, or criminal prevention (e.g., Common EU Anti-Trafficking Plan and European Multidisciplinary Platform Against Criminal Threats). For instance, the EC launched the EU Talent Pool Pilot to facilitate labor integration of Ukrainians who are registered for Temporary Protection in EU countries. In that, the European employment services as a European cooperation network between the EC, the European Labor Authority,

[7] Temporary Protection Directive (Council Directive 2001/55/EC) represents an exceptional tool providing immediate and temporary protection in case of a mass influx of displaced persons from non-EU countries who are not able to return to their country of origin. The TPD was adopted after the conflict in former Yugoslavia.

national public employment services, and other Members and Partners in Europe managed to coordinate a database of over 3 million job vacancies throughout seven EU member states, including Poland (EC, 2022a). To help EU states neighboring Ukraine to manage arrivals at the borders efficiently – reducing the waiting time but still maintaining a high level of security – the EC published guidelines on external border management with Ukraine on March 2, 2022, which also proved very useful.[8] Moreover, the EC regularly updates the guidelines on general visa issuance for Russian citizens at the external border. On September 9, 2022, a full suspension of the Visa Facilitation Agreement with Russia was announced with the aim of putting pressure on the Russian Federation as the aggressor in the Russian–Ukraine war. The EU, however, did provide support not only to its member states but also to its remote periphery. This can be demonstrated by pointing to the EU Support Hub for Internal Security and Border Management in Moldova, which aims to reinforce Moldovan resilience in dealing with the strain stemming from the influx of displaced Ukrainians. This project follows up on the EU–Moldova agreement on border cooperation management with Frontex from March 17, 2022 (EC, 2022c).

We can see that although the EC was at the very center of strengthening the resilience of its peripheral member states at the borders with Ukraine, other EU agencies played an indispensable role as well. Frontex, besides the EU–Moldova agreement, was present on the ground to support first-entry EU countries through two operational activities since January 26, 2022 – Joint Operation Terra and Joint Operation Coordination Points. Europol also deployed operation teams to the frontline European countries neighboring Ukraine, including Poland. The Europol teams support the national authorities with secondary checks and investigations in order to identify potential criminals or terrorists trying to cross the border under false pretenses. Another influential EU agency dealing with Ukraine-related asylum and reception needs in EU countries is the EU Agency for Asylum (EC, 2022c).

The EU also contributed to the fiscal capacity of countries hit by the geopolitics-induced STP. The European Investment Bank (EIB), for instance, dispersed 600 million euros in collaboration with the Polish National Bank (BGK) into the Aid Fund for Ukrainian refugees. The 600 million euros represents only the first tranche of the 2 billion euros to be allocated to support the fund (EIB, 2022). Other EIB assistance also included almost 60 million in grants to repair trains, railways, and bridges in Ukraine and, most importantly, nearly 18 million from the Eastern Partnership Technical Assistance Trust Fund to help communities care for about 700,000 Ukrainians fleeing their homes. Among other countries, this fund was also supported by Poland. Altogether, the EU, member states, and other financial institutions have mobilized over 19.7 billion euros in financial, humanitarian, emergency, and budget support and over 3 billion euros[9] in military assistance under the

[8] Commission Communication Providing operational guidelines for external border management to facilitate border crossings at the EU–Ukraine borders 2022/C 104 I/01.

[9] EU military support for Ukraine provided by the European Peace Facility and the Member States directly is around 9 billion euros.

European Peace Facility[10] to Ukraine since the start of the war. Besides macro-financial assistance and budget support, the grants, loans, and guarantees were also allocated toward building Ukraine's resilience in cyber security (10 million euros), digital transformation (15 million euros), and civil society (31 million euros). Another 18 billion euros is planned for 2023 (EC, 2022b). Although most of the financial aid reallocated from the EU goes directly into Ukraine and therefore reinforces primarily Ukrainian resilience and capacity building, according to our conceptual framework (Fig. 4.7), this will inter alia contribute to the EU's periphery resilience as well. The EU's resilience toward the geopolitics-induced STP was also accentuated by the swift imposition of sanctions against the Russian Federation designed to hamper the Kremlin's ability to wage the war and to impose economic and political costs on Russia's political elite responsible for indication. The sanctions, among other things, contain travel bans, asset freezes, and import bans on Russian coal or seaborne crude oil. The EU's reaction regarding the sanctions has been very quick as the first set of sanctions was agreed already on February 23, 2022, one day before the official invasion was launched. With the aim of politically strengthening the resilience of its periphery, the EU offered EU candidacy status, despite widespread enlargement fatigue among its member states, to both Ukraine and Moldova (Bosse, 2022).

Given all the schemes the state allowed for the municipalities and local authorities, it comes as no surprise that these non-state actors were among the most active in terms of helping the Ukrainian refugees. The Polish municipalities played a significant role in helping the state to govern the abrupt refugee crisis. We have already seen that some cities, such as Krakow, have experienced an increase in population by almost 20%. Moreover, the report published by the Union of Polish Metropolises shows that 69% of Ukrainian refugees in Poland are in the 12 largest cities and metropolitan areas (Wojdat & Cywiński. 2022). Cities and municipalities were at the forefront of this crisis and without any doubt born a heavy burden in terms of budget, equipment, and infrastructure (NDI, 2022). The cities and municipalities provided mainly services, such as housing, education, health care, and public services (e.g., public transport, waste management, water supply, and sewage), most of them free of charge. Under the Act on Public Benefit and Voluntary Organizations, many local authorities proceeded to bidding procedures in an urgent mode in order to provide support for Ukrainian refugees, namely psychological counseling, purchase of beds, and accommodation (Ociepa-Kicińska & Gorzałczyńska-Koczkodaj, 2022). Regarding access to means of subsistence, a one-time allowance of 300 PLN (65€) on food, clothing, footwear, personal hygiene products, and housing costs per person was made available for Ukrainians under the Special Act, whereas the allowance was paid by the municipalities (OECD, 2022). Provision of the material needs, however, was not the only credit the municipalities should be claimed for. They also oversaw the coordination of activities of 14,000 volunteers. The municipalities, in some cases, also coordinated the process of matching offers of private

[10] First time in its history that the EU provided military support to a third country at war.

accommodation to beneficiaries through an electronic system with city services and trained volunteers verifying the proposals of accommodation. As a result, a private person who provides accommodation and food to a Ukrainian citizen might receive a benefit on the basis of a contract with the municipality up to 40 PLN per person per day within a 60-day period (Dobiás & Homem, 2022). The cooperation between various non-state actors and between state and non-state actors can be identified as one of the most important determinants of success when it comes to governing and being resilient against social tipping points. We have already outlined how the state directly supported the municipalities, cities, and NGOs via various funding schemes and legal amendments.

Private actors and NGOs started providing aid to the incoming refugees in the first days following the war's outbreak. This was even before the state managed to mobilize and systemize aid (legal amendments and funding) and when large inflows of refugees were already crossing the Polish–Ukrainian border. Ten thousand private cars decided to pick up the Ukrainian refugees right from the border, and hundreds of thousands of Polish provided private accommodation to the refugees. According to the research of the Polish Economic Institute, 77% of Poles have been directly involved in helping Ukraine refugees spending an estimated 2 billion USD and 7% of the respondents said they even welcomed refugees into their apartments or houses (Baszczak et al., 2022). Ociepa-Kicińska and Gorzałczyńska-Koczkodaj (2022) write about private Polish companies giving Ukrainians jobs and converting empty office spaces into company kindergartens for the children of refugees. Another important aspect of the private help was organization. Since day 1, the NGOs have relied on volunteers working for free, helping the refugees, collecting food, clothing, and other essential articles such as drugs, and organizing donations for Ukrainian people fleeing the war. As we can see, the private actors' response toward these social tipping points was rather vigorous and energetic. We observe non-state private actors, such as private persons, companies of all sizes, and NGOs investing their respective capacities into the aid for the Ukrainian refugees, providing them with administrative, legal, infrastructural, and fiscal capacities. In contrast to Moria, in this case, the Polish state played a rather positive role in promoting and encouraging the private actors to help via various tax and tax deduction schemes. What most certainly helped was that overall, the attitudes of Poles toward Ukrainians are generally positive. Research conducted by Karakiewicz-Krawczyk et al. (2021) shows that Pols tend to believe that the Ukrainians should be supported in their misfortunes, although these attitudes might be skewed by the prevailing anti-Russian sentiments in the population and already quite substantial Ukrainian diaspora in Poland even before the war.

Contrary to the case of Moria, the handling and governance of the refugee inflows from war-torn Ukraine to Poland have been relatively successful so far. Even before the state managed to mobilize its resources first days following the aggrieve attack of the Russian troops on neighboring Ukraine, various non-state stakeholders, namely the EU, local municipalities, and cities as well as private persons, Polish companies, and NGOs started investing in their respective capacities into the overall Polish resilience and complimented the state when some of the effects of the social tipping point soaked through it. The Polish governance of the Ukrainian refugee crisis represents a textbook representation of our conceptual framework displayed in Fig. 4.7. Of course, it is difficult to predict what is going to happen and how will Polish resilience stand against forthcoming abrupt changes. For instance, the question now remains how the state will manage to govern the long-term impacts of the refugee crisis. Today, Poland grapples with accommodating over 800,000 Ukrainian children into its already stretched schooling system as well as finding jobs and permanent accommodation, and these are problems that will not just fade away as long as the war in Ukraine keeps continuing. Based on what we have presented, however, it seems to us that successful governance of this social tipping point consequence in the form of the refugee inflows to Poland is likely to lower the probability of another migration-related social tipping point like the one we witnessed in Moria on the night of September 8 and 9, 2020, due to combined resilience of the state and non-state actors.

References

Abou-Chadi, T., Cohen, D., & Wagner, M. (2022). The centre-right versus the radical right: the role of migration issues and economic grievances. *Journal of Ethnic and Migration Studies, 48*(2), 366–384.

Aggestam, F., & Pülzl, H. (2018). Coordinating the uncoordinated: The EU forest strategy. *Forests, 9*(3).

Ahmen, S., Hasan, M. M., & Kamal, R. (2022). *Russia-Ukraine crisis: The effects on the European Stock Market.* European Financial Management.

Angst, A., Rüegg, R., & Forster, B. (2012). Declining bark beetle densities (Ips typographus, Coleoptera: Scolytinae) from infested Norway spruce stands and possible implications for management. *Psyche: A Journal of Entomology, 2012.*

Arnberger, A., et al. (2018). Visitor preferences for visual changes in bark beetle-impacted forest recreation settings in the United States and Germany. *Environmental Management, 61*(1), 209–223.

Åtland, K. (2020). Destined for deadlock? Russia, Ukraine, and the unfulfilled Minsk agreements. *Post-Soviet Affairs, 36*(2), 122–139.

Bale, J. S., et al. (2002). Herbivory in global climate change research: direct effects of rising temperature on insect herbivores. *Global Change Biology, 8*(1), 1–16.

Baszczak, Ł., et al. (2022). *How Polish society has been helping refugees from Ukraine.* Polish Economic Institute.

BBC. (2021). *Moria migrant camp fire: Four Afghans sentenced to 10 years in jail.* https://www.bbc.com/news/world-europe-57441291

Bosse, G. (2022). Values, rights, and changing interests: The EU's response to the war against Ukraine and the responsibility to protect Europeans. *Contemporary Security Policy, 43*(3), 531–546.

CfHR. (2019). *Greece must urgently transfer asylum seekers from the Aegean islands and improve living conditions in reception facilities.* Commissioner for Human Rights, Council of Europe. 31 October 2019. https://www.coe.int/en/web/commissioner/-/greece-must-urgently-transfer-asylum-seekers-from-the-aegean-islands-and-improve-living-conditions-in-reception-facilities

Clarke, R. (2016). The Donbass in 2014: Ultra-Right threats, working-class revolt, and russian policy responses. *International Critical Though, 6*(4), 534–555.

Crawley, H. (2016). Managing the unmanageable? Understanding Europe's response to the migration crisis. *Human Geography, 9*(2), 13–23.

Crofts, R., et al. (2005). *IUCN Mission to Tatra National Park, Republic of Slovakia*, April 2005. https://www.wolf.sk/files/dokumenty/IUCN_EN_zaverecna_sprava_2005.pdf

David, L., et al. (2022). The Russian invasion of Ukraine and its public health consequences. *Lancet Regional Health – Europe*, No. 15.

den Hertog, L. (2016). *EU budgetary responses to the "refugee crisis" reconfiguring the funding landscape* (CEPS Paper in Liberty and Security in Europe, No. 93), May 2016.

Digidiki, V., & Bhabha, J. (2020). EU migration pact fails to address human rights concerns in Lesvos, Greece. *Health Human Rights, 22*(2), 291–296.

Dimitriadi, A. (2020). Refugees at the gate of Europe. In *Hellenic Foundation for European & Foreign Policy*, Policy Brief 112/2020. https://www.eliamep.gr/wp-content/uploads/2020/04/Policy-brief-Angeliki-Dimitriadi-final-1.pdf

Dobiás, K., & Homem, F. (2022). EU cities and regions welcoming Ukrainian refugees – Mapping multilevel coordination. In *European Committee of the Regions, Commission for Citizenship, Governance, Institutional and External Affairs*. https://cor.europa.eu/en/engage/studies/Documents/Ukrainian%20refugees_study.pdf

Don, A., et al. (2012). No rapid soil carbon loss after a windthrow event in the High Tatra. *Forest Ecology and Management, 276*, 239–246.

EC. (2016a). *Implementing the Common European Asylum System: Commission escalates 8 infringement proceedings (IP/15/6276).* European Commission. https://ec.europa.eu/commission/presscorner/detail/en/IP_15_6276

EC. (2016b). *Proposals for a Council Implementing Decision setting out a recommendation for temporary internal border control in exceptional circumstances putting the overall functioning of the Schengen area at risk.* European Commission. https://eur-lex.europa.eu/legal-content/EN/TXT/?uri=celex:52016PC0275

EC. (2022a). *EU talent pool pilot.* European Commission, EURES. https://eures.ec.europa.eu/eu-talent-pool-pilot_en

EC. (2022b). *Factsheet: The European Union and Ukraine.* European Commission. https://ec.europa.eu/commission/presscorner/detail/en/FS_22_3862

EC. (2022c). *Migration management: Welcoming refugees from Ukraine.* European Commission, Migration and Home Affairs. https://home-affairs.ec.europa.eu/policies/migration-and-asylum/migration-management/migration-management-welcoming-refugees-ukraine_en

EC. (2022d). *Temporary protection.* European Commission, Migration and Home Affairs. https://home-affairs.ec.europa.eu/policies/migration-and-asylum/common-european-asylum-system/temporary-protection_en

EIB. (2022). *The needs of refugees.* European Investment Bank. https://www.eib.org/en/stories/ukrainian-poland-infrastructure-refugees

EP. (2016). On the frontline: The hotspot approach to managing migration. In *European Parliament's Policy Department for Citizens' Rights and Constitutional Affairs.* https://www.europarl.europa.eu/RegData/etudes/STUD/2016/556942/IPOL_STU(2016)556942_EN.pdf

Fleischer, O., et al. (2009). Risk assessment of the Tatra Mountains Forest. In K. Střelcová et al. (Eds.), *Bioclimatology and natural hazards* (pp. 145–154). Springer.

Fleischer, O., et al. (2020). Carbon balance and streamflow at a small catchment scale 10 years after the severe natural disturbance in the Tatra Mts, Slovakia. *Water, 12*(10).

Gordon, E., & Larsen, H. K. (2021). The violent inaction of the state and the camp as site of struggle: The perspectives of humanitarian actors in Moria Camp, Lesvos. *European Journal of International Security, 6*(4), 418–438.

Gov.pl. (2022). Informacja dla uchodźców z Ukrainy. In *Ministerstwo Spraw Wewnętrznych i Administracji*. https://www.gov.pl/web/mswia/informacja-dla-uchodzcow-z-ukrainy

Havašová, M., Ferenčík, J., & Jakuš, R. (2017). Interactions between windthrow, bark beetles and forest management in the Tatra national parks. *Forest Ecology and Management, 391*(1), 349–361.

Hlásny, T., et al. (2019). *Living with bark beetles: Impacts, outlook and management options*. European Forest Institute. https://efi.int/sites/default/files/files/publication-bank/2019/efi_fstp_8_2019_0.pdf

Hlásny, T., et al. (2021). Devastating outbreak of bark beetles in The Czech Republic: Drivers, impacts, and management implications. *Forest Ecology and Management, 490*.

Holzberg, B. (2021). 'Wir schaffen das': Hope and hospitality beyond the humanitarian border. *Journal of Sociology, 57*(3), 743–759.

Illiadou, E. (2019). Safe havens and prison islands: The politics of protection and deterrence of border crossers on Lesvos Island. *Graduate Journal of Social Science, 15*(1), 62–88.

IOM. (2016). *Mixed migration: Flows in the Mediterranean and beyond: Compilation of available data and information 2015*. International Organization for Migration. https://www.iom.int/sites/g/files/tmzbdl486/files/situation_reports/file/Mixed-Flows-Mediterranean-and-Beyond-Compilation-Overview-2015.pdf

ITV. (2020). Female refugees face dire situation after aid workers flee Lesbos. *ITV News*. https://www.itv.com/news/2020-03-07/female-refugees-face-dire-situation-after-aid-workers-flee-lesbos

Jonášová, M., Vávrová, E., & Cudlín, P. (2010). Western Carpathian mountain spruce forest after a windthrow: Natural regeneration in cleared and uncleared areas. *Forest Ecology and Management, 259*(6), 1127–1134.

Karagiannis, E. (2014). The Russian interventions in South Ossetia and Crimea compared: Military performance, legitimacy and goals. *Contemporary Security Policy, 35*(3), 400–420.

Karakiewicz-Krawczyk, K., et al. (2021). The opinions of poles about the need to provide humanitarian aid to refugees from the area covered by the Russian-Ukrainian War. *International Journal of Environmental Research and Public Health, 19*(20).

Karamanidou, L. (2021). Migration, asylum policy and global justice in Greece. In M. Ceccorulli, E. Fassi, & S. Lucarelli (Eds.), *The EU migration system of governance: Justice on the move* (pp. 89–118). Palgrave Macmillan.

Konôpka, B., Šebeň, V., & Merganičová, K. (2021). Forest regeneration patterns differ considerable between sites with and without windthrow wood logging in the High Tatra Mountains. *Forests, 12*(10).

Kovalčík, M. (2018). Efficiency of the Slovak forestry in comparison to other European countries: An application of Data Envelopment Analysis. *Central European Forestry Journal, 64*, 46–54.

Kovalčík, M., Tutka, J., & Sujová, K. (2018). *Dopady ochrany prírody a krajiny na lesnícko-drevársky sector*. Národné lesnícke centrum.

Liadze, I., et al. (2022). *The economic costs of the Russia-Ukraine conflict* (National Institute of Economic and Social Research, NIESR Policy Paper 32, 2 March 2022).

Lidner, M., et al. (2010). Climate change impacts, adaptive capacity, and vulnerability of European forest ecosystems. *Forest Ecology and Management, 259*(4), 698–709.

Makrickiene, E., et al. (2019). From command-and-control to good forest governance: A critical interpretative analysis of Lithuania and Slovakia. *Forest Policy and Economics, 109*.

MARD. (2011). *Od veternej apokalypsy ubehlo sedem rokov*. Ministry of Agriculture and Rural Development. https://www.mpsr.sk/od-veternej-apokalypsy-ubehlo-sedem-rokov/92%2D%2D-5199/

McAuliffe, M., & Ruhs, M. (2017). *World migration report 2018*. International Organization for Migration. https://www.iom.int/sites/g/files/tmzbdl486/files/country/docs/china/r5_world_migration_report_2018_en.pdf

Mezei, P., et al. (2014). Factors influencing the wind-bark beetles' disturbance system in the course of an Ips typographus outbreak in the Tatra Mountains. *Forest Ecology and Management, 312*.

MMP. (2022). 50,000 Lives lost during migration: Analysis of missing migrants project data 2014–2022. In *Missing Migrants Project*. https://missingmigrants.iom.int/sites/g/files/tmzbdl601/files/publication/file/2022%2050k%20deaths.pdf

Mol, C., et al. (2017). Migration aspirations and migration cultures: A case study of Ukrainian migration towards the European Union. *Population, Space and Place, 24*(5), 2131.

Morris, J. L., et al. (2018). Bark beetles as agents of change in social-ecological systems. *Frontiers in Ecology and Environment, 16*(S1), 34–43.

MSF. (2018). Crowded, dangerous and insufficient access to healthcare in Moria. In *Medicins sans frontieres*. 4 May 2018. https://www.msf.org/greece-overcrowded-dangerous-and-insufficient-access-healthcare-moria

MSF. (2019). Greece: Three years of "cruel, inhuman, and cynical" treatment of migrants and refugees. In *Medicins sans frontieres*. https://www.doctorswithoutborders.org/latest/greece-three-years-cruel-inhumane-and-cynical-treatment-migrants-and-refugees

MSF. (2020). Fire in Moria – If the only way to break free is to burn down your house, something is very wrong. In *Medicins sans frontieres*. https://msf-analysis.org/fire-in-moria-if-the-only-way-to-break-free-is-to-burn-down-your-house-something-is-very-wrong/

NDI. (2022). *Poland's cities on the front lines of Russia's War*. National Democratic Institute. https://www.ndi.org/our-stories/polands-cities-front-lines-russia-s-war

Netherer, S., & Schopf, A. (2010). Potential effects of climate change on insect herbivores in European forests – General aspects and the pine processionary moth as specific example. *Forest Ecology and Management, 259*(4).

Neville, D., Sy, S., & Rigon, A. (2016). On the frontline: The hotspot approach to managing migration. In *Directorate-General for Internal Policies*. https://www.europarl.europa.eu/RegData/etudes/STUD/2016/556942/IPOL_STU(2016)556942_EN.pdf

Nikolov, C., et al. (2014). Post-disaster forest management and bark beetle outbreak in Tatra National Park, Slovakia. *Mountain Research and Development, 34*, 326–335.

Ociepa-Kicińska, E., & Gorzałczyńska-Koczkodaj, M. (2022). Forms of aid provided to refugees of the 2022 Russia-Ukraine War: The case of Poland. *International Journal of Environmental Research and Public Health, 19*(12).

OECD. (2022). *Rights and support for Ukrainian Refugees in receiving countries*. OECD Publishing.

Økland, B., et al. (2016). Transition from windfall- to patch-driven outbreak dynamics of the spruce bark beetle *Ips typographus*. *Forest Ecology and Management, 363*, 63–73.

Orhan, E. (2022). The effects of the Russia-Ukraine War on global trade. *Journal of International Trade, Logistics and Law, 8*(1).

Pérez-Sales, P., et al. (2022). Refugee camps as torturing environments – An analysis of the conditions in the Moria Reception Center (Greece) based on the torturing environment scale. *Environmental Research and Public Health, 19*(16).

Person, R., & McFaul, M. (2022). What Putin fears most. *Journal of Democracy, 33*(2).

Petrakos, G., & Psycharis, Y. (2016). The spatial aspects of economic crisis in Greece. *Cambridge Journal of Regions, Economy and Society, 9*(1), 137–152.

Pollozek, S., & Passoth, J. H. (2019). Infrastructure European migration and border control: The logistics of registration and identification at Moria hotspot. *Environment and Planning D: Society and Space, 37*(4).

Psaropoulos, J. (2020). A brief history of Moria. *The Critic*. 29 December 2020. https://thecritic.co.uk/a-brief-history-of-moria/

Rohozinska, J., & Shpak, V. (2019). Ukraine' s Post-Maidan struggles: The rise of an "outsider" president. *Journal of Democracy, 30*(3), 33–47.

Scheffer, A., & Weber, G. (2022). Russia' s war on Ukraine: The EU's geopolitical awakening. In *The German Marshall Fund of the United States*. https://www.gmfus.org/news/russias-war-ukraine-eus-geopolitical-awakening

Schelhaas, M. J., et al. (2010). Assessing risk and adaptation options for fires and windstorms in European forestry. *Mitigation and Adaptation Strategies for Global Change, 15*, 681–701.

Scipioni, M. (2017). Failing forwards in EU migration policy? EU integration after the 2015 asylum and migration crisis. *Journal of European Public Policy, 25*(9), 1357–1375.

Seidl, R., Rammer, W., & Lexer, M. J. (2010). Climate change vulnerability of sustainable forest management in the Eastern Alps. *Climatic Change, 106*, 225–254.

Šimonovičová, A., et al. (2019). Soil microbiota of dystric cambisol in the High Tatra Mountains (Slovakia) after windthrow. *Sustainability, 11*(23).

Singh, A., Patel, R., & Singh, H. (2022). Recalibration of priorities: Investor preference and Russia-Ukraine conflict. *Finance Research Letters, 50.*

ŠOPSR. (2018). Vysoké Tatry 14 rokov po ničivej víchrici. In *Štátna ochrana prírody Slovenskej republiky.* https://www.sopsr.sk/web/?cl=20399

Strzyżowski, D., Fidelus-Orzechowska, J., & Żelazny, M. (2018). Sediment transport by uprooting in the forested part of the Tatra Mountains, southern Poland. *Catena, 160*, 329–338.

Szabó, J., & Jančovič, P. (2020). *Social tipping points & asylum seekers: Fragility in the EU's remote island areas.* In International Conference on European Integration 2020, Ostrava, pp. 694–707.

TRT. (2020). *How the Greek island of Lesbos became a hub of right-wing activism.* TRT World. https://www.trtworld.com/magazine/how-the-greek-island-of-lesbos-became-a-hub-of-right-wing-activism-36275

Unhcr. (2018). *UNHCR urges Greece to address overcrowded reception centers on Aegean islands.* UNHCR. https://www.unhcr.org/news/briefing/2018/8/5b88f5c34/unhcr-urges-greece-address-overcrowded-reception-centres-aegean-islands.html

Vollmer, B. (2016). *Ukrainian migration and the European Union – Dynamics, subjectivity, and politics.* Palgrave Macmillan.

Wojdat, M., & Cywiński, P. (2022). *Urban hospitality: Unprecedented growth, challenges and opportunities: A report on Ukrainian Refugees in the largest Polish cities.* Union of Polish Metropolises. https://metropolie.pl/artykul/34-million-ukrainians-in-poland-new-report-by-the-union-of-polish-metropolises

Zielonka, T., & Malcher, P. (2009). The dynamics of a mountain mixed forest under wind disturbances in the Tatra Mountains, Central Europe – A dendroecological reconstruction. *Canadian Journal of Forest Research, 39*(11).

Chapter 7
Resilience of the EU's Periphery vis-à-vis Social Tipping Points: Policy Recommendations

The previous section presented reports of our three case studies. Based on theoretical concepts developed in Sects. 2 through 5, we attempted to get insights into factors determining the success or failure of social tipping point governance in the European Union (EU)'s eastern and southern periphery. To summarize our conceptualization, we consider social tipping points to be small quantitative changes that trigger a non-linear response in the social component of the social–ecological system, which are subsequently driven by a self-reinforcing positive feedback mechanism and lead to an often irreversible and qualitatively different state of the social system. Social tipping points and various side effects stemming from them represent an unexpected strain on the states and have to be governed. Subsequently, we work with the concepts of governance, state capacity, and state resilience arguing that they remain to be state-centric concepts. Governance is a dynamic process of the general exercise of authority within which the state's capacity embodies the capacity to implement the state's defined goal. The concept of governance and state capacity is, therefore, necessarily intertwined. State capacity has numerous nuances, but we mostly point to five of its aspects – administrative, legal, infrastructural, fiscal, and military – that, in our opinion, construct the backbone of the state capacity. The state capacity is a linkage between governance and resilience since resilience can be viewed as the state's ability to withstand unexpected disturbances and adapt to them, in our case, primarily the side effects from social tipping points. We further hypothesize that although resilience as the capacity to withstand disturbances and adapt to them is first and foremost a state-centric concept, if the state resilience is insufficient vis-à-vis social tipping points and its malign impacts on society and some of its side effects manage to soak through the state umbrella, non-state actors – both national and transnational – have the capacity to intervene with their respective capacities and complement the resilience of the overall society and also indirectly the state from further uncontrollable feedback loops. To this end, we focus on the EU's peripheral countries as the weakest links within the EU's resilience against social tipping points. It is demonstrated that the ability to withstand disturbances from social tipping points and their side effects and adapt to them is a spatial condition and the most vulnerable countries appear in the EU's periphery. Based on this conceptual framework summarized in Sect. 5 (Fig. 5.1), we selected three case studies – Moria refugee camp burndown in the Greek island of Lesvos (southern periphery), the bark beetle outbreak in Slovak Tatra Mountain (eastern periphery), and the war in Ukraine resulting

J. Szabó et al., *Governance of Social Tipping Points*, Key Challenges in Geography,
https://doi.org/10.1007/978-3-031-47413-2_7

in an influx of displaced Ukrainians to Poland (northeastern periphery) – trying to analyze how the EU's periphery manages to govern and is resilient against the social tipping points induced by migration, climate change, and geopolitics.

In the case of the Moria camp burndown, we witnessed how migration and refugee inflows into Europe caused overpopulation on the Greek island of Lesvos and the lack of resilience on behalf of the state in combination with the lack of resilience of other non-state actors resulted in the societal fragility and subsequently into the emergence of another tipping point – Moria camp burndown. It became obvious that even though the resilience of the non-state actors was not entirely weak as the NGOs, for instance, kept trying to save the island from humanitarian disaster, ultimately, insufficient state capacity and resilience pushed the volunteers and the NGOs out of the island and deteriorated the situation even further which culminated into another social tipping point. Similarly, the Tatra Mountain case revealed how a lack of state capacity and resilience vis-à-vis the climate change–induced windstorm triggered a social tipping point with far greater damages, the outbreak of bark beetles. Albeit the state as a primary stakeholder in governing this social tipping point reacted quite promptly at first, weak resilience in governing the social tipping point side effects in combination with the low capacity of other non-state actors, namely the NGOs, dragged the whole society down and resulted in the occurrence of another social tipping points because the windstorm was not governed properly. Lastly, following the Russian aggression in Ukraine beginning on February 24, 2022, this geopolitics-induced social tipping point triggered various side effects influencing the EU's periphery, namely the exodus of displaced Ukrainians and energy price spikes. We demonstrated that the Polish state in coordination with other non-state actors, primarily local municipalities, civil society, the NGOs, and the EU, managed to govern this STP and proved itself resilient, not contributing to the emergence of another social tipping point. Even if potential tipping points occur in the future, the state and non-state actors proved they are resilient against nonlinear disturbances and will be able to mitigate their side effects. The last case study supports the argument of this monograph that even if the state capacity and resilience are missing vis-à-vis social tipping points, non-state actors' resilience can complement the state and enhance the overall societal resilience. This, however, requires necessary cooperation between the state and non-state actors, their mutual unification regarding the goal implementation as well as some level of state resilience secured. If the state resilience were simply too low, as we could witness in the Moria case, even additional non-state resilience and capacity would not have been sufficient.

These three cases, covering social tipping points induced by migration (Moria camp burndown), climate change (Tatra Mountain bark beetle outbreak), and geopolitics (war in Ukraine), tend to verify our hypotheses laid down in Chap. 5. It was verified that successful governance of social tipping points is a function of the state capacity and, therefore, also of the state resilience. We also demonstrated that in case the state capacity and state resilience fall short of the impact stemming from social tipping points, non-state actors (national and transnational) can mobilize and complement the state with their respective capacities and resilience, strengthening the capacity and resilience not only of the society as a whole but also of the state itself. The non-state actors often have the capacity, as was demonstrated in the case of war in Ukraine and subsequent tensions on the Polish–Ukraine border, to avert the emergence of another social tipping point. Otherwise, the unproperly governed social tipping point can trigger a new social tipping point as was the case in Moria and Slovak Tatra Mountains. Based on our conceptual framework and three case studies, we formulate six policy recommendations.

7.1 Do Not Focus on Governing Social Tipping Points, Govern the Side Effects

Social tipping points represent a challenging social phenomenon with potentially malign consequences, whose occurrence will only increase in the future given the rising complexity of society, economy, and politics of the twenty-first century. In the previous chapters, it was accepted that social tipping points represent small qualitative changes that may trigger abrupt, nonlinear changes in the social component of the social–ecological systems. What is intricate is that these changes are driven by a self-reinforcing positive feedback mechanism and therefore tend to lead to a qualitatively different state of the social system with (often) limited reversibility. Multiple states, abruptness, feedback, and limited reversibility form the backbone of social tipping points. What was emphasized throughout this monograph is that social tipping points within the social–ecological system approach represent abrupt (nonlinear) changes with a threshold that is difficult to determine, which makes them essentially unpredictable and in essence ungovernable. For instance, as was demonstrated in the part concerning the Moria camp burndown, the flow of people coming to Europe from Africa and the Middle East has been steadily growing for years before 2015. We explained how climate change, political grievances, and societal upheavals contributed to triggering the Arab Spring and civil war in Syria. All these incremental changes along with the already ongoing migration flows from the less (economically) developed countries in Africa and Asia to Europe have suddenly and abruptly tipped the whole system over into a new, qualitatively different state with unprecedented flows of migrants and refugees into Europe. Similarly, climate change has been leading to numerous socioeconomic upheavals in the last couple of decades; however, it is very difficult to construct an early warning system that would capture all the possible disruptions that climate change can suddenly trigger in social–ecological systems, including the windstorm in the Tatra Mountain (e.g., Alfieri et al., 2012). In spite of having a relatively clear understanding of the climate change effects and potential disasters it can/will bring about, most of the time, the experts merely predict the likelihood of occurrence of certain types of disasters not when such and such event (e.g., deadly windstorm) is going to take place (IPCC, 2022a, b) The same applies to geopolitical tensions on the Ukraine–Russia border and subsequent eruption of war. Although it seems to us, at least ex post, that the development of the relationship between Ukraine and the Russian Federation followed a clear pathway toward this end, it was close to impossible to determine at which point (and even if) the first boot of a Russian soldier crosses the border and starts off an open offensive unprecedented in the twenty-first century in spite of early warnings provided by the US intelligence (Lubold et al., 2022). The peripheral EU countries simply were not able to predict these events either due to their inherent unpredictability (although that is questionable) or due to their lack of capacity to do so. As a result, the peripheral states and their non-state actors should not invest resources into governing the social tipping points per se, but rather govern their consequences and malign side effects, which implies a need for capacity and resilience building. Scholars should be careful and take into account the possibility "that the complexity of social systems inhibits – at least at the moment – our ability to observe tipping dynamics with the methodological tools at our disposal" (Milkoreit, 2023).

What we suggest is to stop focusing on decoding the heuristics in social–ecological systems with utopic attempts of factoring in all possible scenarios. Instead, we encourage the EU peripheral actors, namely the state, to focus more on governing the side effects stemming from social tipping points in order to avoid further snowball effects and the emergence of subsequent social tipping points resulting from ungoverned consequences of the previous one. As was demonstrated in the previous chapter, it was precisely the mishandled governance of the refugee inflows in the reception center on the Greek island of Lesvos (Moria) that caused its burndown. Similarly, the incapacity of state and non-state actors to govern the removal of fallen trees in Tatra Mountain laid down the foundation for the epidemic of bark beetles. We argue that if the state and non-state actors had sufficiently robust capacity (combined) and were resilient against these social tipping points and thus prepared to face and withstand their impacts and possibly adapt to them, it would not have mattered whether they could have been foreseen. Preparedness thus does not equal forecast ability but rather the capacity to govern even unanticipated abrupt changes within the social–ecological system that could have negative spillover effects over numerous areas of the society. We agree with Juncos (2017) who argues that the capacity to be resilient and to be prepared for unknown risks should be achieved through adaptation, learning by doing, and flexibility rather than trying to eliminate uncertainty. Such uncertain social tipping points tend to emerge no matter what, especially in increasingly complex environments such as the economy, politics, and climate of the twenty-first century with close interlinks among the subsystems as the complexity and disembeddedness of contemporary social–ecological systems appear to be on unprecedented levels (Polanyi, 2001).

Currently, given the highly unpredictable nature of the COVID-19 crisis that has exposed a myriad of vulnerabilities in the EU member states, even the Commission placed resilience at the forefront of its policy goals, not prevention and early warning systems. Thereby, resilience became a "new compass for EU policies" (EC, 2020). We were able to observe that even after the first, relatively mild COVID wave tamped down and the element of surprise disappeared, many countries could still not face the disturbances and adapt to them in the subsequent waves of COVID due to insufficient capacities, especially in the EU periphery (e.g., Hale et al., 2021). Therefore, the EU peripheral countries should also proceed in line with the

Commission's recommendations and focus on capacity and resilience building rather than investing in prevention and early warning systems. The modern history of the EU's periphery reveals that these social tipping points tend to happen no matter what and it is, therefore, necessary to invest in resilience building against the social tipping points, rather than trying to come up with a precise location of their thresholds. In the end, paradoxically, the prevention capacities could also be enhanced with rising capacities in administration, legal and fiscal systems, infrastructure, and military.

Naturally, this recommendation does not aim to encourage policymakers and pivotal stakeholders from overlooking social tipping points altogether, nor does it invoke boundless nihilism and laissez-faire crisis management, on the contrary. In fact, targeting the long-term causes of poverty, climate change or geopolitical tensions are crucial for diminishing the likelihood of catastrophic scenarios with potentially nonlinear turbulence, such as the massive exodus of people from Africa to Europe, pest epidemic, or the war on our eastern borders. However, in our conceptualization, these processes belong to capacity and resilience building, not social tipping point management. Pooing resources into green transition and mitigation of climate change should not be conceptually confused with governing climate change as a trigger for windstorms having adverse and unpredictable effects on the local economy surrounding the Tatra Mountain in Slovakia. Instead, the necessity of capacity building with regard to climate change, governing the windstorm itself and governing the social tipping points side effects such as the removal of infected trees ought to be conceptually separated. Similarly, when a social tipping point such as the war in Ukraine happens, it should be a priority to govern the sudden energy spikes and refugee inflows, not the war itself, especially not by the EU peripheral countries that are economically, politically, and ecologically constrained given their geography and size. All in all, this recommendation aims to suggest that one should be "careful when making claims about how the world is, that is, the occurrence of social tipping points, and their observability with scientific methods and data" (Milkoreit, 2023).

7.2 Strengthening the State Capacity

For every state to successfully govern, including the EU peripheral states, the states need to be able to define their goals, reconcile them, implement the goals, and finally get feedback to secure the learning curve. In all of this, the capacity to implement the state's goals – the state capacity – is the most crucial one, especially when facing social tipping points. In the end, it is the state capacity that determines the overall state resilience or the ability of the state to withstand the disturbances stemming from social tipping points and subsequently adapt to them. It is therefore quite straightforward. In order to be more resilient against social tipping points, society has to invest and strengthen the capacity of the state as the state remains the primary driver of governance in society. Strengthening the state capacity requires investing in respective aspects of the state capacity: (1) administrative, (2) legal, (3) infrastructural, (4) fiscal, and (5) military.

It was emphasized multiple times throughout this monograph, but the modern state cannot function and implement its goals without a proper bureaucratic apparatus as policies envisaged by the state will not be attained without administrative employees of the state ready to perform all the tasks entrusted to them. The existence of professional and politically insulated Weberian bureaucracies is therefore a necessity in twenty-first-century societies, especially in democratic societies such as the ones in the EU's periphery (Cingolani et al., 2015). For the EU's peripheral states to be able to implement necessary policies when facing social tipping points (induced by migration, climate change, or geopolitics), the existence of bureaucracy with sufficiently high enough human capital is a must. For instance, the burndown of Moria camp was to a certain extent caused by bureaucratic incapacity to administer the asylum requests, failure to manage the basic functioning of the camp by the Greek authorities, or lack of administrative information for the asylum seekers. Investing in the bureaucracy will necessarily augment the state capacity and resilience when facing abrupt changes, which has been proved by numerous authors (e.g., Bruszt & Vukov, 2017). However, one cannot emphasize enough the need to increase not only the quantity of the bureaucracy but also the quality, especially when upcoming social tipping points in the twenty-first century will take up more and more sophisticated forms, such as the onset of AI, technological disruption, or massive unemployment caused by automation. Bureaucratic capacity relying on sufficiently enough human capital will become priceless. Thus, it is necessary for peripheral states to invest in capacities enhancing the human capital of its bureaucratic apparatus through partnerships and scholarships between leading national universities and the state, to envisage programs to attract young people studying abroad to come back to their countries of origin and work for the state and also to improve wage but also nonwage conditions for the white-collar workers in the services of the state in order to attract the most qualified people. Even if it does not seem to be that important, improving nonwage conditions via various work-related benefits such as home offices or day cares and fulfilling certain sociocultural needs such as decent discourse and respecting LGBTQ+ rights might be a game changer for attracting people with higher education. Furthermore, administrative capacity building is not merely about the personal substrate. Based on the OECD recommendations, good governance practices, strategic planning and coordination, and institutional building are similarly, if not even more important than personal capacities and should be, therefore, implemented as well (OECD, 2020). The peripheral EU states would be better off by implementing good administrative governance practices already well-functioning in the western and northern parts of the EU.

Administrative capacity is closely linked to legal capacity or the state's ability to enforce contracts and property rights, sustain the rule of law, and administer the court and prison system. Proper functioning of the legal system, namely judicial dispute settling, and protection of fundamental rights and liberties are crucial in reinforcing resilience against social tipping points which is especially visible in migration-related crises as transparent and, more importantly, prompt judicial systems are important features when dealing with disputes over asylum requests during refugee crises. This is a place where administrative and legal capacity necessary overlap. Furthermore, not only protection but also socially acceptable elasticity of property rights and environmental laws is also a necessary condition for an adequately functioning legal system. This was evident in the case of fallen trees in Tatra Mountain that were left on the ground in the protected forest areas. It was forbidden by the law to remove the fallen trees from the protected areas in Tatra Mountain even though it was absolutely necessary to do so in order to avoid the bark beetle epidemic. If the legal capacity of the state was of sufficient quality, the corresponding property and protection elasticity would react sooner. Among other things, the legal state apparatus is also indispensable in terms of the right to legal aid. We can see that the peripheral countries of the EU are less likely to provide affordable and accessible aid in legal matters, be it civil, criminal, or administrative (World Justice Project, 2022). Affordability and accessibility of the fundamental right to legal aid have to be secured for the proper functioning of every legal system.[1] Reallocation of resources into the legal capacity of the state and its effective utilization is thereby necessary to strengthen the EU's periphery resilience; otherwise, a potential social tipping point might one day unexpectedly overwhelm the whole legal system in these countries with citizens in social need unable to access the justice system and claim their rights.

In line with the Commission recommendations,[2] we are convinced that in order to become more resilient, more resources have to be allocated to critical infrastructure capacity belonging to the state. This was revealed drastically in the context of the war in Ukraine (physical and cyber-attacks, hybrid threats, sabotage of the Nord Stream gas pipelines, and transport of weaponry), the COVID-19 pandemic, and the European refugee crisis (Duarte et al., 2022). The Moria reception center is a suitable example of how malfunctioning and insufficient infrastructure – in terms of hygiene, shelter, and access to health professionals – can cause irreversible damage to the refugee community and trigger a humanitarian catastrophe. It is, therefore, self-evident that the administrative and legal apparatus of the state cannot function properly if the infrastructure is of insufficient quality and quantity. Strengthening of the infrastructural capacity is therefore a necessary condition of resilience enhancement, primarily in countries of the EU's periphery suffering from infrastructural debt. The state should have at its disposal infrastructure to support its defined goals, their subsequent implementation, and resilience. Naturally, there are dozens of possible infrastructural projects that should be supported in order to strengthen the state resilience. At this point, in order to avoid an exhaustive list of infrastructures that we think should be supported, it is important to emphasize the digital and energy infrastructure in our opinion. According to the Commission's plans for the Digital

[1] Article 6 (3)(c) of the European Convention on Human Rights and Article 47 of the Charter of Fundamental Rights of the European Union.

[2] See, for instance, the proposal for a directive of the European Parliament and of the Council on the resilience of critical entities (2022/0365 COD), the Council Directive 2008/114/EC, or the European Parliament and Council Directive 2015/1148.

Decade, if Europe wants to achieve digital leadership, it needs to invest in secure and sustainable digital infrastructure for connectivity, microelectronics, and data processing (EC, 2021). The EU's ambition is that all EU households will have gigabit connectivity, and all populated areas will be covered by 5G Internet connection, whereas the EU is planned to secure 20% of world sustainable semiconductors production and deploy 10,000 climate-neutral edge nodes in the EU. The peripheral EU countries need to take more initiative in this area. It comes as no surprise that digital infrastructure is of national importance for these countries given their fragility vis-à-vis automation and hybrid threats (e.g., Luptáčik et al., 2021), and this fragility will only gain importance concerning the current onset of AI. Furthermore, the infrastructure regarding energy resilience needs to be strengthened and the EU peripheral countries should take ownership of this topic, especially in the east. Although the EU made great progress in reducing the dependency on Russian gas and oil following the Russian aggression in Ukraine, the necessary energy infrastructure is still lacking. In 2021, EU countries imported 155 billion cubic meters of gas, 108.1 million tons of crude oil, 91 million tons of petroleum products, and 51.4 million tons of coal from Russia, whereas the Russian Federation also provided fuel for the operation of 18 nuclear blocks in the EU, of which most of the nuclear plans were in the peripheral EU countries (Czechia, Hungary, Slovakia, Finland, and Bulgaria). In 2022, Russian gas imports to the EU dropped by 80 billion cubic meters and the EU imposed an embargo on the import of Russian coal, seaborne crude oil, and petroleum products. Overall supply of oil from Russia to the EU decreased by 90% (Kardaś, 2023). Nevertheless, the security of supply remains very fragile in the eastern peripheral countries (e.g., EC, 2022b). It is thus imperative, in line with the seventh state of the EU Energy Union report, to "accelerate the transition to clean energy and bring dependence on Russian energy to an end as soon as possible and well before the end of this decade" (EC, 2022c). Based on the Commission's estimations, a fully fledged reduction of dependence on Russian fossil fuels will require 300 billion euros and the peripheral EU countries are in acute need of pushing forward the EU agenda of energy infrastructure building according to the EU energy union goals aiming for sustainable and green future (EC, 2022a).

Neither the administrative-legal nor infrastructural capacity can be built if the state lacks financial resources to create, enhance, or maintain them. In the twenty-first century, fiscal capacity entails the state's capacity not only to raise revenues or to borrow money from lenders but also to technocratically concentrate fiscal resources into areas according to the specified and defined goals and at the same time not jeopardize the country's fiscal sustainability. This requires skillful administrative staff in state treasuries, political leadership, and even constitutional tools to avert unnecessary spending. Greece, an example of a highly indebted economy with multiple experiences of state defaults in its modern history, may be symptomatic of its inability to raise revenues (tax avoidance), forestall fiscal wastage (corruption and state capture), and redirect fiscal resources into targeted areas (malfunctioning technocratic apparatus). This eventually causes a lack of investment into resilience in various aspects, ranging from administration to infrastructure and military. Moria reception center can be again a very good example of that. The lack of fiscal capacity corruption-wise, however, does not concern only Greece but is typical of all EU peripheral countries, including Slovakia and Poland (e.g., Transparency, 2022). Furthermore, in the context of the euro debt crisis, it became obvious that the current architecture of the EU's economic governance is closely linked to the fiscal capacities of its member states. It is thus important for the peripheral EU countries to push for a better accommodation of their economic models into the EU economic governance with its current biases. It is argued by numerous authors that the EU economic governance after the debt crisis openly prioritizes export-led growth models functioning on the same principles as the German manufacturing-goods exporting economy, whereas the growth models reliant upon domestic consumption are explicitly penalized (Johnston & Matthijs, 2022). This directly concerns the southern euro peripheral countries – Greece, Spain, Italy, and Portugal – as it puts constraints on the fiscal capacities of their respective countries and by that account on their ability to enhance their resilience vis-à-vis social tipping points.

Lastly, all three analyzed cases in Chap. 6 – Moria reception center, Tatra Mountain bark beetle outbreak, and war in Ukraine – demonstrate the necessity to invest in capacity and ipso facto resilience in military and law enforcement. The military apparatus, which in our understanding covers all law enforcing units (including police, intelligence services, and even firefighters) with proper equipment and training, allows the state to govern a wide range of consequences emerging from social tipping points. Be it a need to organize and sustain migration flows crossing the Mediterranean, help the foresters remove the fallen trees, or assist the displaced people at the Polish–Ukrainian borders, it is mostly force and intervention units that handle these disturbances. Allocation of resources into training, equipment, and personal growth in these law-and-order enforcement units is another important step in increasing resilience in the EU's periphery.

7.3 Investing in Non-state Actors' Resilience

For the most part, authors tend to focus primarily on the state when assessing the concept of resilience and its role in governing societal changes, such as globalization (e.g., Fjäder, 2014). We do not oppose this approach, on the contrary. Within our conceptual framework, we uphold the state-centric approach in resilience against social tipping points since the state remains and is expected to remain on the frontline when social tipping point consequences and side effects, such as migration, bark beetle outbreak, or energy spikes, emerge. The state's ability to withstand disturbances and to reorganize and adapt afterward is therefore crucial. However, in line with recent research (e.g., Sellberg et al., 2021; Bodin et al., 2022), we also call for multi-actor collaboration to address disturbances uprising from abrupt changes in social–ecological systems. According to our conceptual framework, which was also demonstrated in our three case studies, a lack of non-state resilience can also be to the detriment of not only the society but the state itself. Complementary to our recommendation to strengthen the state capacity, we also encourage the state itself and pivotal stakeholders to invest in non-state actors' capacity and resilience as such "interventions may contribute actively to the resilience of the overall system, by enhancing the entities' own abilities to cope with disturbances" (Manca et al., 2017). It is also important beyond simple crisis management as the determinants that enhance a region's resilience during disturbances tend to average be the same that enhance their growth potential and competitiveness during stable times (e.g., Fratesi & Rodrìguez-Pose, 2016; Di Caro & Fratesi, 2018).

When we refer to non-state resilience, we think of a construct of sums of fractional resiliencies of the non-state actors, both national (local and private) and transnational (EU and the others) that is, in turn, determined by their respective capacities (administrative, legal, infrastructural, fiscal, and military). In the end, the cumulation of state and non-state resilience defines the overall resilience of the society and by definition its capacity to withstand disturbances from social tipping points and adapt to them. Subsequently, if societal resilience is insufficient relative to the magnitude of consequences from social tipping points, even the fragility of the state increases, which makes the state dependent on the capacity and resilience of the non-state actors. We, therefore, encourage society, and in particular the state, to start investing in non-state actors' resilience if it wants to remain or become more immune to nonlinear changes, such as migration, climate change, or geopolitics. Sufficiently high enough resilience of non-state actors in the case of Polish–Ukraine borders, namely the EU and local actors (Polish municipalities), supported the state and ipso facto its resilience against the unexpected strain having its origin in Russian aggression in Ukraine. Investing in the non-state actors' resilience in Poland, namely municipalities and NGOs, might help prevent further occurrence of social tipping points (e.g., Bristov & Healy, 2020). For instance, the non-governed side effects (displaced Ukrainians) from the geopolitics-induced social tipping point may cause another social tipping point and trigger a snowball effect. One can easily imagine how the rapidly growing Ukrainian population in Polish cities may contribute to increasing social tensions related to the situation in the labor market, housing crisis, deterioration of social services, and in the long run even potential political cleavages, in turn, influencing the state resilience (Wojdat & Cywiński, 2022). In general, it is thus argued that the reallocation of sufficient resources into non-state actors' resilience building, such as the EU, NGOs and nonprofit organizations (NPOs), municipalities and other local self-governments, and even private companies tied to R&D and universities will strengthen not only the overall societal resilience but also the state's resilience (see also Anholt & Wagner, 2019). Albeit it might seem such a self-evident observation, the non-state actors remain to be overlooked when it comes to societal resilience building. For instance, within the EU's flagship regarding the resilience building – Next Generation EU package – there is still very limited involvement of local and regional governments in the design of the national recovery and *resilience* plans (CoR, 2022), not to mention other aspects of the civil society, such as NGOs, private companies, and local volunteers.

A problematic aspect of investing in non-state actors' resilience might be a reluctance on the part of the general public and civil society to channel public resources into non-state stakeholders without proper oversight over these finances. Although we do not claim that the investments have to be necessarily of public origin, the concern is not entirely nonsensical. Within the EU's periphery, among countries with relatively well-established democratic institutions and active civil society (shackled European Leviathan),[3] accountability and transparency must be secured, especially since the feedback mechanisms are crucial for the whole process of governance in democratic countries. As a result, the channeling of public resources into non-state actors with the aim of increasing their resilience has to be accompanied by democratic accountability and transparency, which will, in turn, encourage public opinion toward such investments. Especially in the context of local (municipalities and cities) and transnational (EU) stakeholders, the democratic mechanisms must be strengthened for the general public to politically accept further reallocation of fiscal resources into these non-state actors.

7.4 Building More Resilient European Union

Striving to build a more resilient EU periphery will also require building a more resilient Europe in our opinion (see also Juncos, 2017), which is fully in line with the EC'ṣ goals (EC, 2020). As the EU's periphery is facing the Russian aggressor close to its border (namely Baltics states and Poland), ongoing migratory flows through the Mediterranean and from Turkey (Italy and Greece), and impacts of climate change, it will need more European integration and closer ties with the EU, not less. Our conviction comes from the analysis of the role of the EU in (co)governing the geopolitics-induced social tipping point and the side effects from it. We have shown very prompt and effective help from the EU toward its peripheral countries, mainly Poland, which helped the Polish state manage the strain on its borders with Ukraine. On the contrary, there are also examples of the EU's inability to face exogenous shocks, such as COVID-19 due to its insufficient resilience (Barbier-Gauchard et al., 2021). Thereby, we can see that the EU can be a very important stakeholder when it comes to resilience against social tipping points; however, European institutions' architecture has to allow for that. The EU has already undergone numerous attempts to voice the need to fortify its resilience, especially within the European foreign policy circles (e.g., Tocci, 2019) and long-term economic plans (EU Regulation 2021/241),[4] but the incentives have to come first and foremost from the nation states and in our cases from the EU peripheral states. This becomes increasingly difficult as many peripheral EU states, namely self-proclaimed illiberal governments in Hungary and Poland, escalate the tensions with the EC over crucial competencies in resilience infrastructure, such as the energy market and control over consumer prices (Szabo & Fabok, 2020). We agree with the authors that while numerous antagonisms persist between the EU and

[3] In spite of illiberal and antidemocratic tendencies in Hungary and Poland (e.g., Szabó, 2022), certain requirements regarding accountability and transparency are still present in these countries.

[4] EU Regulation 2021/241 of the European Parliament and of the Council of February 12, 2021, establishing the Recovery and Resilience Facility.

national decision-makers, their position is not entirely irreconcilable. Paradoxically, social tipping points and side effects from them, e.g., shortages of gas in Hungary due to state paternalism in energy policies in the context of the war in Ukraine, might help to accelerate such transfer of competencies onto the EU, although the argument that deeper EU integration is forged through crisis might not be as convincing as one might think (Anghel & Jones, 2022).

Once a consensus regarding the transfer of competencies to the EU is achieved, building European resilience will once again require enhancement in the EU's respective capacities. Possibilities are endless, but we would like to emphasize the strengthening of at least three capacities in this place: infrastructural, fiscal, and military capacity. In terms of the infrastructural capacity, we would recommend a greater focus on the industrial policies of the EU, especially shifting European industrial policies in such a way as to enhance the EU's infrastructural capacities. Industrial policies represent an area in which each EU country tends to follow its national strategy. On the contrary, even the role of the EU in shaping national industrial strategies had not been entirely negligible (e.g., Medve-Bálint, 2014). And since there were trends of gradual deindustrialization amid increasing competition coming from China even before the COVID-19 pandemic (Aiginger & Rodrik, 2020), discussions regarding promotion and greater unification of the European industrial policies were at a core of concerns in European economic policy discussions. We agree with Barbier-Gauchard et al. (2021) that the EC policy goals of improving cost competitiveness through constraining the growth of wage costs, essentially internal devaluation, were not the best strategy and more focus should be put on industrial policies in which unit labor costs are improved due to new productivity increases. This, however, also implies reforming the European system of innovation as a whole. According to Pianta et al. (2020), stimulating growth in innovative and sustainable sectors could increase unit labor cost with a positive effect on public–private cohabitation – indirect nonmonetary investment into other non-state actors – and subsequently also for the overall European infrastructural resilience. European projects such as smart grids for photovoltaic parks in Europe could serve as an example (Zsyman et al., 2012) and contribute to the increased resilience in the EU's periphery.

These changes in European industrial policies, however, cannot be achieved under current fiscal and competition rules (Pianta et al., 2020). Regarding the fiscal capacity, we, therefore, recommend changing the fiscal rules in the EU (not only the Eurozone) that should be designed in a way to take into account the requirements of its (peripheral) member states, however, also taking into account the EU's goals regarding green transition (EC, 2019). More emphasis should, therefore, be placed on support for long-term growth rather than shortsighted economic stabilization which means getting rid of the export constraints put on some peripheral EU countries (Johnston & Matthijs, 2022). Furthermore, the process of resilience building should be exploited in favor of more European integration, such as the joint financing of major long-term investment programs or even a common debt instrument (Herzog, 2020). The establishment of the Recovery and Resilience Facility[5] is a

[5]EU Regulation 2021/241 of the European Parliament and of the Council of February 12, 2021, establishing the Recovery and Resilience Facility.

proper step forward and seems to represent the onset of the EU fiscal capacity, however, still not of sufficient size and with question marks regarding the permanence of this mechanism (Fabbrini, 2022). The peripheral EU countries thus need to push toward permanent EU fiscal capacity that would enable the EU represented by the Commission to pool resources on behalf of the EU under more convenient conditions than it is now under asymmetric economic integration (Ibidem). The pooled European resources can be subsequently directed into infrastructural projects, industrial capacities, and other capacities enhancing European resilience, such as digital and energy union resilience. This way, "an economy that works for people" which is built on social inclusion and reduced inequalities can be attained. This will, however, also require a more compact and integrated EU capital market to channel savings to those who need them the most in order to empower small and medium companies which represent the backbone of the EU's economic engine (EC, n.d.).

The war in Ukraine also revealed the need to enhance the EU in its military capacity, especially when it comes to foreign policy.[6] The EC has been obsessed with the concept of resilience in foreign policy for some time now (Juncos, 2017). For instance, the EU High Representative for Foreign Affairs Mogherini officially presented the EU Global Strategy for Foreign and Security Policy (EUGS) already in 2016. The EUGS contains numerous references to the term resilience in EU foreign policy.[7] The term "resilience" even managed to get into the EUGS priorities as building "state and societal resilience to our East and South" was identified as the second of the five key priorities for the EU's external action (EUGS, 2016).[8] In this priority, the EU committed to promoting resilience also in its surrounding regions, regions close to its periphery, as enhancing external resilience automatically strengthens the internal resilience back within the EU (Tocci, 2019). According to the EUGS, a "resilient state is a secure state, and security is key for prosperity and democracy." This will, however, require a more credible enlargement policy as a strategic investment in Europe's security and resilience in our opinion, not deceiving the Western Balkan countries with vague promises of potential EU membership (Petrovic & Tzifakis, 2021). We are convinced that enlargement policies should not only be more predictable, but the enlargement requirements could also contain conditionality regarding resilience building, e.g., capacity in migration, energy security, terrorism, or organized crime. Furthermore, a reversal of the trend of strengthening national powers with the support of EU institutions in the context of the war in Ukraine would be in place (Genschel, 2022). We are aware of the acute need to strengthen the EU military presence, especially in the context of cases such as the Moria reception center burndown or the war in Ukraine. Enhancing the military capacity of the EU does not have to be solely in the form of pooling common resources into the European army (e.g., Kucera, 2019). We

---

[6] As was argued before (Chap. 3), military capacity in our understanding represents a broader concept, covering not only military forces but also law enforcement.

[7] The term resilience is mentioned in the EUGS no less than 41 times in 60 pages (Juncos, 2017).

[8] Alongside (1) building the Union's own security; (2) pursuing an integrated approach to conflicts and crises; (3) supporting cooperative regional orders; and (4) a commitment to a reformed multilateral, rules-based system of global governance.

recommend either strengthening national cooperation in law-enforcement units at the EU level, potentially with a separate EU budget, or creating standby resilience units prepared to help increase the capacity of the states in the EU's periphery to govern various side effects from unexpected social tipping points.

The last point we would like to make regarding our recommendation to build a more resilient Europe is also the necessity to target the democratic deficit and overall quality of democracy at the EU level when facing unexpected crises. Some research has already accentuated the acceleration of the adoption of extraordinary measures and the lack of consultation and transparency during crises (e.g., Rhinard, 2019), and there are also clear tendencies of coercive enforcement at the expense of voluntary cooperation, mainly in the field of fiscal governance (Leontitsis & Ladi, 2018). Likewise, Moury et al. (2021) emphasize the strengthening of non-majoritarian EU bodies such as the European Central Bank, Eurogroup, the European Court of Justice, or Frontex and, in reverse, the weakening of national parliaments across the EU. Moreover, in the context of the European debt crisis, the role of parliaments in debtor countries (the EU's periphery) has been weakened compared with parliaments in creditors' countries (Steinbach, 2019). We can, therefore, observe the gradual dismantling of space for citizens to be included in the decision-making process following the multiple crises, which only tends to alienate the electorate from transferring more competencies onto the EU and by the same token strengthen the overall EU's resilience. If the democratic deficit at the EU level is not addressed, the more resilient Europe will become less likely to attain. Furthermore, the emergence of the next social tipping point (mainly of endogenous economic origin) might only accelerate the de-democratization and even worsen this tension. Building a more resilient Europe will require not only enhancement of the EU's capacity but also political capital invested in the EU's democratization and transparency.

7.5 Fortifying Peripheral Areas as a Precondition for Resilient European Union

In the preceding section, it was argued that the peripheral EU countries should aim to reinforce the EU's capacity and resilience in order to enhance their own capacities and resilience. We, however, argue this to be a two-way street. Not only is the capacity of the peripheral EU countries enhanced by the increase in capacity and resilience of the EU, but the EU of which the peripheral states are a part, especially in the context of integrated core state powers, will also benefit from fortified periphery, especially given its relative economic backwardness when compared with the EU average. Both the EU and its periphery are dependent on each other with respect to resilience against social tipping points.

A glance at the distribution of per capita income across the EU regions in Fig. 7.1 demonstrates that the EU peripheral countries, especially in the east and south, evince the lowest economic development throughout the EU indicating low capacity

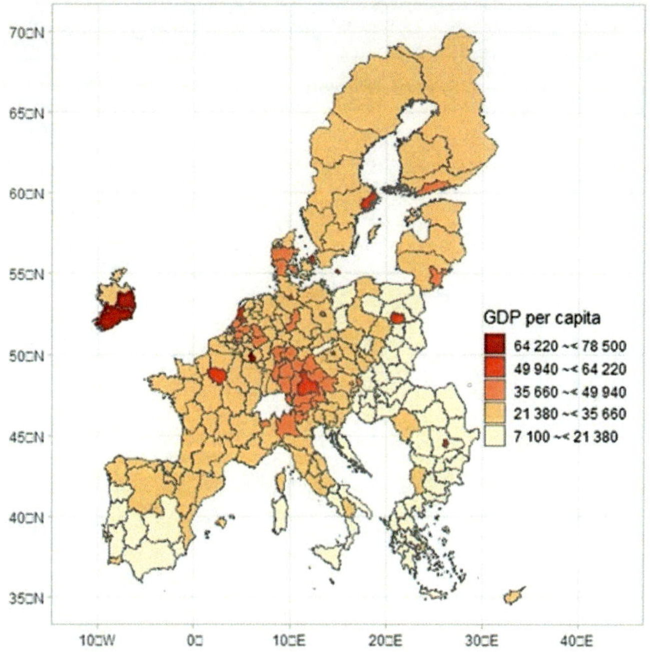

***Source*: Own elaboration. Map produced in R with data Eurostat data (NAMA_10R_2GDP) from Eurostat-package http://ropengov.github.io/eurostat**
***Note*: GDP in purchasing power standard (EU27 = 2020).**

Fig. 7.1 Gross domestic product at current market prices, NUTS2 regions 2020. (*Source*: Own elaboration. Map produced in R with data Eurostat data (NAMA_10R_2GDP) from Eurostat-package http://ropengov.github.io/eurostat)
Note: **GDP in purchasing power standard (EU27 = 2020)**

and subsequently low resilience of these peripheral regions (see also Sect. 4.1). Subsequently, it comes as no surprise that countries in the peripheral regions happen to be well below 100% of the EU27 average GDP per capita (Table 7.1). Therefore, most "least developed" and "transition" regions happen to be in these peripheral countries (Krausova & Walsh, 2022). Given the already demonstrated correlation between the economic standard of living and the degree of vulnerability in Fig. 4.5, it is very important for the EU to invest and transfer resources into the economic development of the peripheral regions that will then spill over into its capacity and resilience building.

Historically, the cohesion policy had been the EU's main investment policy that deliberately tried to target all economically suboptimal regions with the aim of boosting their business competitiveness, economic growth, and sustainable development. The cohesion policy is currently primarily channeled through four funds: European Regional Development Fund (ERDF), Cohesion Fund, European Social Fund Plus (ESF+), and Just Transition Fund which are all under the umbrella of the Investment for Jobs and Growth goal and for the current Multiannual framework

Table 7.1 GDP per capita at market prices across EU countries, 2020

Country	*% of EU27 average*	*GDP per capita at market prices*
Luxembourg	261.5	78,528.2
Ireland	205.1	61,599.0
Denmark	132.7	39,857.0
Netherlands	130.5	39,185.3
Austria	124.9	37,507.0
Germany	123.2	36,998.4
Sweden	122.4	36,774.4
Belgium	119.0	35,728.8
Finland	114.1	34,255.0
France	104.7	31,430.5
EU27 (from 2020)	100.0	30,032.6
Malta	97.9	29,414.8
Italy	94.1	28,261.4
Czechia	93.4	28,047.5
Cyprus	90.4	27,137.7
Slovenia	89.1	26,767.9
Lithuania	87.6	26,317.2
Estonia	86.1	25,848.3
Spain	82.9	24,889.0
Portugal	76.2	22,897.4
Poland	76.1	22,858.8
Hungary	74.5	22,382.7
Romania	72.7	21,831.8
Latvia	71.9	21,603.2
Slovakia	71.8	21,558.7
Croatia	64.8	19,460.3
Greece	62.0	18,614.0
Bulgaria	55.2	16,570.1

Source: Own elaboration based on Eurostat data (NAMA_10_PC)
Note: GDP in purchasing power standard (EU27 = 2020), current prices.
Shaded area as a reference point refers to the EU average values.

(2021–2027) have at disposal more than 360 billion euros (Krausova & Walsh, 2022). The *Cohesion Fund* with its 36.6 billion euros is specifically designed to invest in less prosperous EU countries. In the past, the cohesion policies of the EU indeed contributed to the closing gap between (some) peripheral EU countries and regions and its average (Maynou et al., 2016; Radvanský, 2016). As we can see from Fig. 7.2, the eastern peripheral countries that accessed the EU after 2004 managed

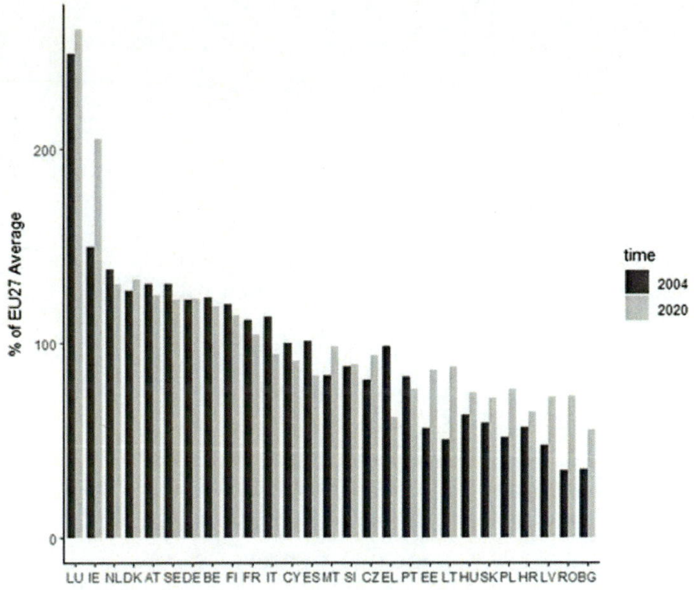

Source: Own elaboration. Figure produced in R with data Eurostat data (NAMA_10_PC) from Eurostat-package http://ropengov.github.io/eurostat
Note: GDP in purchasing power standard (EU27 = 2020), current prices.

Fig. 7.2 EU member states' percentage of EU27 GDP per capita. (*Source*: Own elaboration. Figure produced in R with data Eurostat data (NAMA_10_PC) from Eurostat-package http://ropengov.github.io/eurostat)
Note: **GDP in purchasing power standard (EU27 = 2020), current prices**

to economically converge toward the EU average, whereas the southern peripheral countries, namely Greece, worsened their position from 98.2% in 2004 to just 62% in 2020. Similarly, Spain also dropped by more than 18 percentage points. The relative deterioration of the southern periphery is largely attributed to the European debt crisis (Johnston & Matthijs, 2022) and, therefore, does not result from cohesion policies. Subsequently, it comes as no surprise that the eligibility of investment for jobs and growth goals within the Multiannual framework (2021–2027) almost perfectly overlaps the less developed and in-transition NUTS2 regions in the EU (Fig. 7.3), where the "less developed" regions constitute regions with GDP per capita (purchasing power standard) having less than 75% of the EU27 average and "transition" regions are those with GDP per capita (purchasing power standard) between 75% and 100% of the EU27 average.

The goal of "reducing disparities between the levels of development of the various regions and the backwardness of the least favored regions" is explicitly enshrined in the treaties (TFEU art. 174–178), and the tools of the cohesion policies seem to be, therefore, properly devised. However, we would argue that a little more ought to be done regarding enhancing the economic capacity of the EU peripheral regions as the

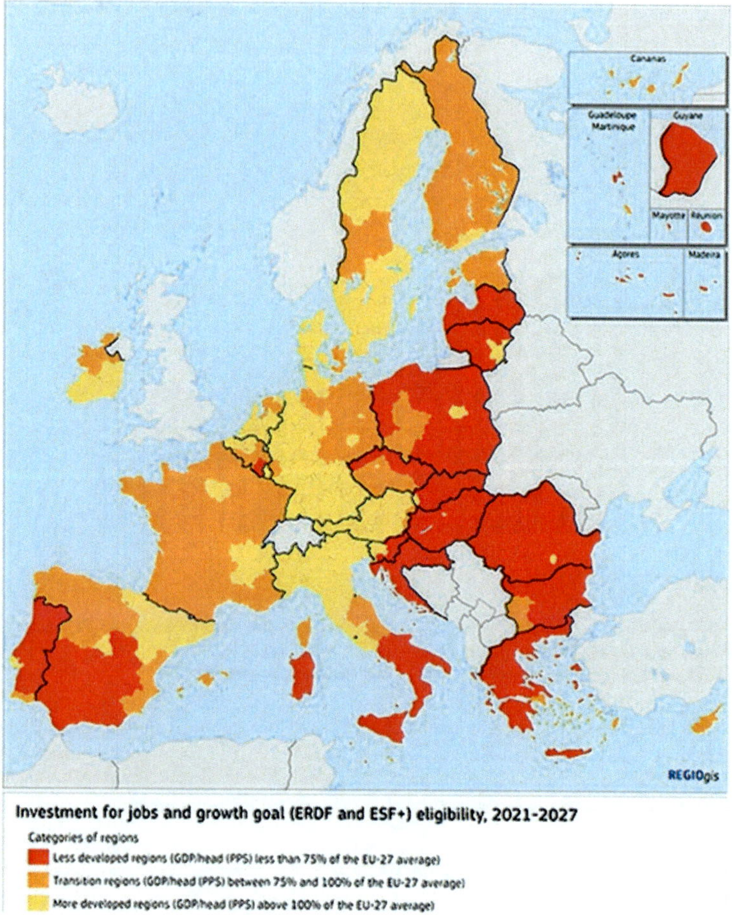

Source: https://ec.europa.eu/regional_policy/information-sources/maps_en#1.

Fig. 7.3 Investment for jobs and growth goal eligibility, 2021–2027. (*Source*: https://ec.europa.eu/regional_policy/information-sources/maps_en#1)

"catching up" process with the EU average remains insufficient, which subsequently mirrors itself in the low capacity and resilience of the EU periphery vis-à-vis social tipping points. For instance, in spite of the eligibility to draw funding from ERDF (313 billion euros along with ESF+) or cohesion funds (36.6 billion euros), the actual allocation is often very cumbersome given the low capacity of the peripheral member states to effectively spend the eligible funds (e.g., Hudec, 2023). The regional absorption capacity of the EU funds is indeed conditional on regional economic development (Kersan-Škabić & Tijanić, 2017). With the vision of strengthening its periphery, the EU represented by the Commission could start devising an asymmetric approach that would allow the underspending peripheral regions to spend the EU funds more effectively. Alternatively, the EU gets stuck in a loop where low-performing peripheral countries are unable to effectively spend the eligible EU funds that are designed to make them more competitive and level their economic development with the EU average which will require further EU funds in the future. In the end, even though the distribution of EU funds does not appear to stimulate support for the EU, a combination of effective use of the available EU funds and proper management might enhance the EU image (López-Bazo, 2022).

Another recommendation might be a greater deployment of the European Investment Bank (EIB) as the EU "lending arm" in promoting regional economic convergence. Again, despite being heavily under-researched (e.g., Clifton et al., 2018), the EIB remains to be the largest multilateral financial institution and largest provider of green finance which earned it the title "climate bank" (EIB, 2023a) with assets exceeding 600 billion euros (Mertens & Thiemann, 2018). Although some of its activities are of global outreach, most financing programs are tied to the EU as the shareholders of the EIB are the EU member states who are fully eligible for financing operations (their share in the bank's capital is based on their economic weight within the EU at the time of accession). Thus, one of the EIB's priorities is also to support cohesion and balanced territorial development as it finances support projects in EU regions below the EU GDP per capita average, mostly with regard to innovation, infrastructure, small- and medium-sized enterprises, and environmental sustainability (EIB, 2021). In a nutshell, the EIB complements the EU funds with loans and other financial instruments according to its priorities regarding cohesion. In the period 2014–2020, it provided more than 120 billion euros in EU cohesion regions. Furthermore, the European Investment Fund, as part of the EIB group, plays an important role in co-managing EU funds within EU cohesion policy on behalf of national and regional authorities (Ibidem). We perceive a possibility of the EIB's role in increasing EU cohesion in two aspects. First, the EIB's advisory services already provide technical support and strategic expertise to regional and national authorities and financial intermediaries. However, closer cooperation between the Commission, the EIB advisory capacity, and national and regional authorities regarding project management and project implementation might allow greater absorption of the EU funds in peripheral EU regions, reinforcing the EU itself. Such cooperation is already underway in supporting the member states in the national Recovery and Resilience plans (EIB, 2023b). This will necessarily require deploying more human capital of the EIB into EU cohesion priorities. Second, the EIB is already closely cooperating with the Commission within Juncker's Investment Plan for Europe and its core piece European Fund for Strategic Investment. The plan also encompasses the European Investment Advisory Hub and the European Investment Project Portal that are expected to offer advisory information for potential investors (Mertens & Thiemann, 2018). On the contrary, we agree with Mertens and Thiemann (2018) that the EIB's participation in a collation of supranational institutions, national governments, development banks, and private financial actors in reshaping the European bank-based system of corporate governance into a market-based finance system might have gone too far and can have detrimental effects on the EU political economy. We would, therefore, call for a more cautious approach and recommend the EIB as a development bank to focus primarily on promoting regional cohesion, diminishing economic disparities, and providing expertise to the EU member states in need, fully in line with the European Green Deal agenda, and stick to its agenda of "promoting regional economic integration and development over the alleviation of capital constraints" (Clifton et al., 2018). This can be conducted through a greater involvement of the EIB in issuing common European debt instruments that will be backed by the ECB and the funds subsequently channeled into the EU peripheral regions with the aim of enhancing their economic capacities and resilience.

7.6 Rethinking Methodological Approach Toward Social–Ecological Systems

It was pointed out multiple times throughout this monograph that disregarding the ecological aspect might be detrimental to resolving the most pressing social tipping points in the twenty-first century. The IPCC (2022a, b) reports are very clear about this. Climate change brought about by irresponsible actions of the human population (e.g., emission of CO2) causes various malign climate-related phenomena (droughts, floods, and extreme weather) and has adverse effects on numerous aspects of society, economy, and politics. For instance, it is well established that climate change has an impact as direct, through the destruction of shelters and food insecurity, as well as indirect, through drops in rural income losses and health problems, drivers of involuntary migration and displacement (Ibidem). It was also shown that climate change and extreme climate events might even lead to violent conflicts, such as the one in Syria in March 2011, culminating in immense migratory strain on Europe with numerous societal consequences. Indirectly, climate change through violent conflicts and migration influences labor markets, housing, and even parliamentary elections. As a case of the social tipping point induced by climate change, we have pointed to the windstorm in 2004 that caused a massive bark beetle outbreak in the Slovak Tatra Mountain. The windstorm was caused to a large extent by climate change–related phenomena and culminated in 2.8 mil m^3 of forest uprooted. Subsequently, as 165,000 m^3 of fallen wood was left uncleared, a new social tipping point emerged from the side effects of the previous one – the bark beetle outbreak – having far-reaching impacts not only on forest management but local economy relying on timber processing as well, not to forget slump in tourism or increase in health-related problems. Naturally, social and environmental dynamics are intertwined even in reverse order, with public policies directly contributing to the deterioration of the environment (e.g., backing coal-intensive industries in Poland; Szabo & Fabok, 2020).

Environmental and societal problems are closely intertwined, and it is thus methodologically inappropriate to separate them. Our last policy recommendation is, therefore, to urge policymakers and academics to try to adopt an interdisciplinary social–ecological approach when assessing social tipping points, state resilience, and non-state actors' capacity whenever it becomes possible. Although some of them are still of very important academic contribution (e.g., Alessi et al., 2020), an analysis of the resilience within the broader social–ecological context intertwined with an interdisciplinary approach would benefit all, the policymakers, academics, and, most importantly, the public. Methodologically speaking, the policymakers and academics should thus switch from a solely social system approach into a social–ecological system approach in order to be able to holistically estimate the overall need to build, enhance, or maintain state and non-state resilience against social tipping points resulting not only from migration, climate change, and geopolitics but also from unforeseeable nonlinear shifts such as the onset of the AI, technological disruptions, or automation. This is one of the reasons why, within our methodological framework, we proceeded with Milkoreit et al.'s (2018) conceptualization of social tipping points with a social–ecological system as our base. Social tipping points within the context of the social–ecological system were subsequently defined as the threshold at which "a small quantitative change inevitably triggers a nonlinear change in the social component of the social–ecological system, driven by self-reinforcing positive feedback mechanisms, that inevitably and often irreversibly lead to a qualitatively different state of the social system" (Milkoreit et al., 2018: 10). Given the interconnectedness between social and ecological system components, the crossing of a social tipping point leads to a qualitatively different social–ecological system. It is time for us to realize that neither the social nor the ecological systems are present in a vacuum and should be, therefore, analyzed jointly whenever the situation and circumstances allow for that.

References

Aiginger, K., & Rodrik, D. (2020). Rebirth of industrial policy and an agenda for the twenty-first century. *Journal of Industry, Competition and Trade, 20*(2), 189–207.

Alessi, L., et al. (2020). The resilience of the EU member states to the financial and economic crisis. *Social Indicators Research, 148*, 569–598.

Alfieri, L., et al. (2012). Operational early warning systems for water-related hazards in Europe. *Environmental Science & Policy, 21*, 35–49.

Anghel, V., & Jones, E. (2022). Is Europe really forged through crisis? Pandemic EU and the Russia – Ukraine war. *Journal of European Public Policy*.

Anholt, R., & Wagner, W. (2019). Resilience in the European Union external action. In E. Cusumano & S. Hofmaier (Eds.), *Projecting resilience across the Mediterranean* (pp. 17–36).

Barbier- Gauchard, A., et al. (2021). Towards a more resilient European Union after the COVID-19 crisis. *Eurasian Economic Review, 11*, 321–348.

Bodin, Ö., et al. (2022). Choose your collaborators wisely: Addressing interdependent tasks through collaboration in responding to wildfire disasters. *Public Administration Review, 82*(6), 1154–1167.

Bristov, G., & Healy, A. (2020). *Handbook on regional economic resilience*. Edward Elgar Publishing.

Bruszt, L., & Vukov, V. (2017). Making states for the single market: European integration and the reshaping of economic states in the Southern and Eastern peripheries of Europe. *West European Politics, 40*(4), 663–687.

Cingolani, L., Thomsson, K., & de Crombrugghe, D. (2015). Minding Weber more than ever? The impacts of state capacity and bureaucratic autonomy on development goals. *World Development, 72*, 191–207.

Clifton, J., Díaz-Fuentes, D., & Gómez, A. L. (2018). The European Investment Bank: Development, integration, investment? *JCMS: Journal of Common Market Studies, 56*(4), 733–750.

CoR. (2022). *EU annual report on the state of regions and cities 2022*. European Committee of the Regions. https://cor.europa.eu/en/our-work/Pages/State-of-Regions-and-Cities-2022.aspx

Di Caro, P., & Fratesi, U. (2018). Regional determinants of economic resilience. *The Annals of Regional Science, 60*, 235–240.

Duarte, A. P., Redzepagic, S., & Murta, F. S. (2022). *The European integration process: Crisis and resilience in the aftermath of the COVID-19 pandemic*. Coimbra University Press.

EC. (2019). *The European green deal*. European Commission. https://eur-lex.europa.eu/legal-content/EN/TXT/?qid=1576150542719&uri=COM%3A2019%3A640%3AFIN

EC. (2020). *Strategic foresight report*. European Commission. https://commission.europa.eu/strategy-and-policy/strategic-planning/strategic-foresight/2020-strategic-foresight-report_en

EC. (2021). *2030 Digital compass: The European way for the digital decade*. European Commission. https://commission.europa.eu/strategy-and-policy/priorities-2019-2024/europe-fit-digital-age/europes-digital-decade-digital-targets-2030_en

EC. (2022a). *A framework strategy for a resilient energy union with a forward-looking climate change policy*. European Commission. https://eur-lex.europa.eu/legal-content/EN/TXT/?uri=COM:2015:80:FIN

EC. (2022b). *Slovakia energy snapshot 2022*. European Commission, State of the energy union 2022 – Snapshots per EU country. https://energy.ec.europa.eu/document/download/51661e08-322d-46b5-b3b4-7516c346338a_en?filename=SK_2022_Energy_Snapshot.pdf

EC. (2022c). *State of the energy union 2022*. European Commission. https://eur-lex.europa.eu/legal-content/EN/TXT/?uri=CELEX%3A52022DC0547&qid=1666595113558#footnote57

EC. (n.d.). *An economy that works for people: Ensuring social fairness and prosperity*. European Commission. https://commission.europa.eu/strategy-and-policy/priorities-2019-2024/economy-works-people_en

EIB. (2021). *Cohesion and regional development overview*. European Investment Bank. https://www.eib.org/en/publications/cohesion-and-regional-development-overview-2021.htm

EIB. (2023a). *EIB at glance*. European Investment Bank. https://www.eib.org/en/publications/20220244-eib-at-a-glance.htm

EIB. (2023b). *The EUB Group operational plan, 2023-2025*. https://www.eib.org/attachments/lucalli/20220289_eib_group_operational_plan_2023_en.pdf

EUGS. (2016). *Shared vision, common action: A stronger Europe. A global strategy for the European Union's foreign and security policy*. European Union Global Strategy. https://www.eeas.europa.eu/sites/default/files/eugs_review_web_0.pdf

Fabbrini, F. (2022). *EU fiscal capacity: Legal integration after Covid-19 and the war in Ukraine*. Oxford University Press.

Fjäder, C. (2014). The nation-state, national security and resilience in the age of globalization. *Resilience, 2*(2), 114–129.

Fratesi, U., & Rodríguez-Pose, A. (2016). The crisis and regional employment in Europe: what role for sheltered economies? *Cambridge Journal of Regions, Economy and Society, 9*(1), 33–57.

Genschel, P. (2022). Bellicist integration? The war in Ukraine, the European Union and core state powers. *Journal of European Public Policy, 29*(12), 1885–1900.

Hale, T., et al. (2021). A global panel database of pandemic policies (Oxford COVID-19 government response tracker). *Nature Human Behaviour, 5*, 529–538.

Herzog, B. (2020). Whither coronabonds? The past and future of the EMU in the Coronavirus Pandemic. *Intereconomics, 55*(3), 155–158.

Hudec, M. (2023). Slovakia among worst in spending EU funds. *Euractiv*. https://www.euractiv.com/section/politics/news/slovakia-among-worst-in-spending-eu-funds/.

IPCC. (2022a). *Climate Change 2022: Impacts, adaptation and vulnerability*. Working Group II contribution to the IPCC Sixth Assessment report of the intergovernmental panel on climate change. https://www.ipcc.ch/report/ar6/wg2/downloads/report/IPCC_AR6_WGII_FullReport.pdf

IPCC. (2022b). *Climate Change 2022: Mitigation of climate change*. Working Group III contribution to the IPCC Sixth Assessment report of the intergovernmental panel on climate change. https://www.ipcc.ch/report/ar6/wg3/downloads/report/IPCC_AR6_WGIII_Full_Report.pdf

Johnston, A., & Matthijs, M. (2022). The political economy of the Eurozone's post-crisis growth model. In L. Baccaro, M. Blyth, & J. Pontusson (Eds.), *Diminishing returns: The new politics of growth and stagnation* (pp. 117–142). Oxford University Press.

Juncos, A. E. (2017). Resilience as the new EU-foreign policy paradigm: A pragmatist turn? *European Security, 26*(1), 1–18.

Kardaś, S. (2023). Conscious uncoupling: Europeans' Russian gas challenge in 2023. In *European Council on foreign relations*. https://ecfr.eu/article/conscious-uncoupling-europeans-russian-gas-challenge-in-2023/#:~:text=Before%20Russia%20invaded%20Ukraine%2C%20the,cent%20of%20total%20gas%20imports

Kersan-Škabič, I., & Tijanić, L. (2017). Regional absorption capacity of EU funds. *Economic Research-Ekonomska Istraživanja, 30*(1), 1191–1208.

Krausova, T., & Walsh, J. (2022). *2021–2027: Cohesion policy EU budget allocations*. European Commission. https://cohesiondata.ec.europa.eu/stories/s/2021-2027-EU-allocations-available-for-programming/2w8s-ci3y/

Kucera, T. (2019). What European army? Alliance, security community or postnational federation. *International Politics, 56*, 321–338.

Leontitsis, V., & Ladi, S. (2018). The changing nature of European governance and the dynamics of Europeanization. In E. Ongaro & S. Van Thiel (Eds.), *The Palgrave handbook of public administration and management in Europe* (pp. 765–781).

López-Bazo, E. (2022). The impact of cohesion policy on regional differences in support for the European Union. *JCMS: Journal of Common Market Studies, 60*(5), 1219–1236.

Lubold, G., Gordon, M. R., & Trofimov, Y. (2022). U.S. Warns of imminent Russian invasion of Ukraine with tanks, jet fighters, cyberattacks. *Wall Street Journal*. https://www.wsj.com/articles/ukraine-troops-told-to-exercise-restraint-to-avoid-provoking-russian-invasion-11645185631

Luptáčik, M., et al. (2021). *Hospodárske a sociálne dopady priemyslu 4.0*. VEDA SAV.

Manca, A. R., Benczur, P., & Giovannini, E. (2017). *Building a scientific narrative towards a more resilient EU Society*. European Commission, JCR Science For Policy Report. https://publications.jrc.ec.europa.eu/repository/bitstream/JRC106265/jrc106265_100417_resilience_scienceforpolicyreport.pdf

Maynou, L., et al. (2016). The impact of structural and cohesion funds on Eurozone convergence, 1990-2010. *Regional Studies, 50*(7), 1127–1139.

Medve-Bálint, G. (2014). The role of the EU in shaping FDI flows to East Central Europe. *Journal of Common Market Studies, 52*(1), 35–51.

Merten, D., & Thiemann, M. (2018). Market-based but state-led: The role of public development banks in shaping market-based finance in the European Union. *Competition & Change, 22*(2), 184–204.

Milkoreit, M. (2023). Social tipping points everywhere? Patterns and risks of overuse. *WIREs Climate Change, 14*(2).

Milkoreit, M., et al. (2018). Defining tipping points for social-ecological systems scholarship – An interdisciplinary literature review. *Environmental Research Letters, 13*, 1–13.

Moury, C., et al. (2021). *Capitalising on constraint: Bailout politics in Eurozone countries*. Manchester University Press.

OECD. (2020). *Strengthening governance of EU funds under cohesion policy: Administrative capacity building roadmaps*. OECD. https://www.oecd.org/publications/strengthening-governance-of-eu-funds-under-cohesion-policy-9b71c8d8-en.htm

Petrovic, M., & Tzifakis, N. (2021). A geopolitical turn to EU enlargement, or another postponement? An introduction. *Journal of Contemporary European Studies, 29*(2), 157–168.

Pianta, M., Lucchese, M., & Nascia, L. (2020). The policy space for a novel industrial policy in Europe. *Industrial and Corporate Change, 29*(3), 779–795.

Polanyi, K. (2001). *The great transformation: The political and economic origins of our times.* Beacon Press.

Radvanský, M. (2016). *Impact of cohesion policy on regional development of Slovakia: Ex-post assessment of national strategic reference framework 2007–2013.* Institute of Economic Research, SAS. https://ekonom.sav.sk/uploads/journals/334_impact_of_coheion_policy_on_regional_development_of_slovakia.pdf

Rhinard, M. (2019). The crisification of policy-making in the European Union. *Journal of Common Market Studies, 57*(3), 616–633.

Sellberg, M., et al. (2021). Engaging with complexity in resilience practice. *Ecology and Society, 26*(3).

Steinbach, A. (2019). EU economic governance after the crisis: revisiting the accountability shift in EU economic governance. *Journal of European Public Policy, 26*(9), 1354–1372.

Szabó, J. (2022). Political economy od illiberal capitalism in Hungary and Poland. *Politická ekonomie, 70*(5), 617–637.

Szabo, J., & Fabok, M. (2020). Infrastructure and state-building: Comparing the energy politics of the European Commission with the governments of Hungary and Poland. *Energy Policy, 138.*

Tocci, N. (2019). Resilience and the role of the European Union in the world. *Contemporary Security Policy, 41*(2), 176–194.

Transparency. (2022). *Corruption Perceptions Index 2021.* Transparency. https://images.transparencycdn.org/images/CPI2021_Report_EN-web.pdf

Wojdat, M. & Cywiński, P. (2022). *Urban hospitality: Unprecedented growth, challenges and opportunities: A report on Ukrainian Refugees in the largest Polish cities.* Union of Polish Metropolises. Available at https://metropolie.pl/artykul/34-million-ukrainians-in-poland-new-report-by-the-union-of-polish-metropolises

World Justice Project. (2022). *Rule of law index.* World Justice Project. https://worldjusticeproject.org/rule-of-law-index/downloads/WJPIndex2022.pdf

Zsyman, J., et al. (2012). Green growth. *Intereconomics, 47*(3), 140–164.

Chapter 8
Conclusion

In times when resilience in social and economic, geopolitical, green, and digital dimensions was introduced as an official compass for EU policymakers (EC, 2020a), the concept of resilience as the ability to withstand and cope with endogenous as well as exogenous challenges while undergoing transitions in a sustainable, fair, and democratic manner, secured its prominence in the EU policymaking. The COVID-19 pandemic has undoubtedly unfolded the EU's need to enhance its resilience. This becomes especially important for the EU's southern and eastern periphery as well, arguably the most fragile parts of the EU. The ambition of this monograph was thus to fill the gap in the literature, which tends to overlook the EU's peripheral countries in the east and south in terms of their resilience building against the abrupt nonlinear changes erupting in one system (e.g., climate) but having impacts beyond the system's boundaries (e.g., society, economy, and politics). In order to analyze the resilience of the peripheral EU countries, we employed an interdisciplinary approach, combining theoretical intersections from the social tipping point literature, public policy, regional economic resilience, and European studies. This allowed us to shed light on two, still relatively understudied areas in the resilience literature – social tipping points and EU peripheral studies – and present a conceptual framework of governance of tipping points in the EU periphery. The conceptual framework links together the concepts of social tipping points, governance, state capacity, state resilience, and interactions between the state and non-state actors. Moreover, the multilevel governance approach is also complemented with a coherent description of the structural factors contributing to lower levels of resilience to the detriment of the EU periphery in the east and south. Subsequently, the conceptual framework is tested against three case studies, each social tipping point with a different trigger (migration, climate change, and geopolitics) and each social tipping point situated in a different EU peripheral country (Greece, Slovakia, and Poland). Such a multiple case study research design ensures greater robustness and validity of the results.

Based on the case study reports, the main takeaways of the monograph can be summarized as follows. First, the state with its capacity is crucial for a successful

J. Szabó et al., *Governance of Social Tipping Points*, Key Challenges in Geography,
https://doi.org/10.1007/978-3-031-47413-2_8

governance of social tipping points. In our conceptualization, the state stands at the forefront of resilience and acts as a protective shield over its polity. Be it Greek islands in the Aegean Sea defended by the coastguard, protected woodlands in Northern Slovakia overseen by the state institutions, or Polish regions neighboring Ukraine with the border police in standby regime, the state was always hit first by these unexpected events. The state capacity in all of its forms is, therefore, crucial for the successful governance of unexpected disturbances. Even if the state was not ready to face side effects stemming from social tipping points, we still live in a state-centric world with a centralized monopoly on the use of force and prima facie authority of the state. Thus, even in the context of the erosion of the state within the globalized world, the state remains primarily responsible for the protection of its citizens against social tipping points and other disturbances and its potential ability to do so is heavily dependent on its administrative, legal fiscal, infrastructural, and military capacity. The second conclusion concerns a situation when the state capacity simply falls short and is of insufficient quality to protect its citizens. Within the multilevel governance framework, non-state actors can step in and complement the state with their respective capacities. Furthermore, the successful governance of social tipping points requires congruence in the state actors' and non-state actors' goals. We have seen how insufficient cooperation between the state and non-state actors failed to prevent burning of the Camp Moria on the Greek island of Lesvos and how the antagonistic perception of the social–ecological threats stemming from the fallen trees in the protected forest areas in Slovakia between the state on one hand, and the Slovak environmental NGOs on the other, caused the emergence of pest epidemic having even more far-reaching malign consequences than the original climate-induced windstorm. Hence, in the context of insufficient capacity and resilience of the state, non-state actors can prevent the state and the society in which they live from collapsing if their respective capacities and resilience complement the state and if their goals with regard to governance are in congruence with the state's goals. The third important takeaway concerns the geographical constraints imposed by the distribution of resilience capacities across the EU. As was suggested on multiple occasions throughout this monograph, the eastern and southern periphery attain lower levels of resilience than their western and northern counterparts. This relative lack of resilience seems to be caused by a combination of compositional, collective, and contextual factors. Albeit long suggested in the literature, a comprehensive study of these factors supported by supplementary and descriptive quantitative data allows us to conclude the EU needs to invest in resilience building in its eastern and southern periphery as a stronger periphery is also in the interest of the EU itself as well as its western and northern member states.

Based on these takeaways, this monograph suggests six policy recommendations that may help the policymakers in the EU periphery to govern social tipping points, nonlinear changes in social–ecological systems, more successfully. First, we argue that crucial stakeholders should not try to govern the social tipping points themselves but rather the side effects and consequences stemming from them. This recommendation is based on the assumption that tipping points are mostly unpredictable, and it is always better to be resilient against potential disturbances of all kinds than wait for abrupt changes that might never arise. Second, we encourage policymakers to keep building the state capacity, in all five of its aspects – administrative, legal, infrastructural, fiscal, and military – and this way keep enhancing the state's resilience. Third, we strongly encourage the policymakers to keep strengthening the non-state actors' resilience as well as to keep building cooperation between state and non-state actors. Although the state is a primary stakeholder in governing social tipping points within our conceptual framework, non-state actors tend to step in once side effects from unexpected disturbances soak through the umbrella of state resilience and complement the state in its overall resilience. Not only does society benefit from this additional capacity and resilience, but so does the state itself when a loop of follow-up social tipping points is averted. Fourth, we are convinced that in order to make countries within the EU periphery more resilient against tipping points, the EU has to become more resilient, mainly in fiscal, infrastructural, and military aspects. A necessary condition for building a more resilient EU is, however, a process of mitigating the democratic deficit within the tangle of the EU institutions. Fifth, given the distribution of the resilience capacities throughout the EU as explained in Sect. 4, not only does the periphery need to increase the EU's capacity so do the EU and the core member states need to start dealing with enhancing the resilience in the EU's eastern and southern periphery. In the end, it will benefit both sides. Lastly, we are of the opinion that policymakers should try taking into account the interconnectedness between social and ecological systems and consider them linked whenever possible. This methodological shift toward adopting a social–ecological system approach as a methodological starting point will allow the pivotal stakeholder to make more complex decisions regarding resilience building against social tipping points.

In combination with the stated policy recommendations, this monograph attempted to shed some light on the resilience of the eastern and southern periphery of the EU. Albeit a humble attempt, we hope this monograph can spark a debate on how less developed peripheral countries, in a relative term, govern and can enhance their resilience against tipping points within the social–ecological systems. This debate becomes especially important as the EU and countries of its eastern and southern periphery are expected to face numerous challenges, upheavals, and crises in the years and decades to come. Some are already known, such as the onset of AI, climate change–related social disruptions, automation and its impact on the labor market, or security threats. Some, however, are currently too complex and so far off our current understanding of the current world that cannot be foreseen from today's perspective. The question now is how should the research outlined in this monograph proceed in this context. There are several ways in our opinion. We strongly encourage the researchers to employ a transdisciplinary approach, combining theoretical and empirical knowledge from economics, economic geography, political science, sociology, social psychology, international relations, as well as earth and environmental studies. This way, the intradisciplinary analysis allows for more robust and comprehensive research. Reductionist views on the issue of resilience can lead to different outcomes in the field of regional economics and in the climate change literature. Given that the twenty-first century is likely witnessing unexpected disturbances in both social and ecological systems simultaneously, we need input from both ends. In addition, this also concerns the methodological tools employed. Neither qualitative nor quantitative research alone can provide robust and reliable results based on which relevant policy recommendations can be made. As we have hopefully shown, they need to be used intertwiningly. Most appropriately, by using qualitative research (e.g., case studies or content analysis) supported by empirical data or quantitative research methods (e.g., multilevel structural equation modeling or spatial regression analysis) complemented by supplementary narrative analysis. Furthermore, this also requires the collection of more reliable and accessible data for social and environmental scientists. While the initial attempts of the Commission are a very good starting point for an analysis of resilience within the EU context (EC, 2020b), a more complex database is needed. One that would include other areas of resilience as well, such as health, social trust, education, and institutional quality. Such data may be additionally supplemented with survey data enquiring EU citizens about their preferences and concerns regarding resilience, for instance as part of special eurobarometers. Lastly, this might be one of the most ambitious research challenges lying ahead, and the multilevel governance of social tipping points would benefit from supporting data as well. This means that structural interviews and regular surveys inquire into the interactions between the state and non-state actors, non-state actors' material as well as immaterial resources and the available capacities of state institutions.

References

EC. (2020a). 2020 strategic foresight report: Charting the course towards a more resilience Europe. *European Commission*. Available at https://commission.europa.eu/strategy-and-policy/strategic-planning/strategic-foresight/2020-strategic-foresight-report_en#documents.

EC. (2020b). Resilience dashboards. *European Commission*. Available at https://commission.europa.eu/strategy-and-policy/strategic-planning/strategic-foresight/2020-strategic-foresight-report/resilience-dashboards_en.